Speaking Stata Graphics:
A Collection from the Stata Journal

Speaking Stata Graphics:
A Collection from the Stata Journal

NICHOLAS J. COX
Durham University
Department of Geography

A Stata Press Publication
StataCorp LP
College Station, Texas

Copyright © 2014 by StataCorp LP
All rights reserved. First edition 2014

Published by Stata Press, 4905 Lakeway Drive, College Station, Texas 77845
Typeset in LaTeX 2_ε
Printed in the United States of America
10 9 8 7 6 5 4 3 2 1

ISBN-10: 1-59718-144-7
ISBN-13: 978-1-59718-144-0

Library of Congress Control Number: 2014936700

Contents

Preface   vii

Software notes on the original columns   xiii

Graphing distributions 1

Graphing categorical and compositional data 24

Graphing agreement and disagreement 50

Graphing model diagnostics 71

Density probability plots 98

The protean quantile plot 113

Smoothing in various directions 132

Graphs for all seasons 152

Turning over a new leaf 175

Spineplots and their kin 196

Between tables and graphs 213

Creating and varying box plots 234

Creating and varying box plots: Correction 253

Paired, parallel, or profile plots for changes, correlations, and other comparisons 256

The statsby strategy 275

Graphing subsets 284

Transforming the time axis 296

Axis practice, or what goes where on a graph 306

Trimming to taste 319

Preface

The *Stata Technical Bulletin* was started in 1991 as a journal by and for users of the statistical software Stata. As a Stata enthusiast, I subscribed from issue 1, started contributing in 1994, and became an associate editor in 1998. In 2001, a decision was made to transform the *Bulletin* into the *Stata Journal*, and I became one of the editors. The initial meeting setting up the journal was a working breakfast in a Boston hotel, otherwise made memorable by my being offered cottage cheese as an alternative to yogurt with my cereal. Among many other details, two suggestions followed in quick succession: that there should be a column in the *Journal* and that I should write it. I remain gratefully aware of the compliment implied.

I settled on a title, *Speaking Stata*, to match both the privilege of sounding off on my own Stata-linked preoccupations and prejudices and the purpose of explaining the use of Stata's language in a way that grew from, but went beyond, the excellent Stata documentation. As a columnist, I claim no more than certain interests and points of view; I take comfort in the simple thought that others are free to write on quite different topics and from quite different perspectives.

This book collects those columns published between 2004 and 2013 with a major theme of statistical graphics. My strong interest in graphics has grown steadily out of my education and experience as an academic geographer and a long-standing fascination with geometry and spatial patterns. Despite greatly renewed attention given to graphics over the last few decades within statistical science, I think it remains true that they are generally undervalued within several disciplines, but this is not the place to argue that view in any detail. The columns also include coverage of several little-known and even some apparently new ways of graphing data. A keen personal interest in the history of ideas, which surfaces in several columns, underlines the extent to which good ideas are persistently reinvented, so one can never be sure.

I have received strong and consistent counsel not to attempt any serious rewriting. The columns were certainly written one at a time and are intended to be read, mostly, one at a time. Unlike Dickens, Dumas, or Dostoyevsky, I have not been writing a novel in serial installments. Nevertheless, my own interests should give some coherence to the collection.

Learning statistics software, like learning statistics itself, can be a series of small struggles until enlightenment dawns and you wonder why on earth the idea appeared difficult in the first place. While writing these columns, I have tried to be clear rather than clever because I know from my own experience that several ideas in Stata, although beautifully simple when considered in the right way, are far from transparent at first sight. Similarly, some repetitions remain in the hope that they provide reinforcement or serve the purposes of those with neither time nor inclination to read every column.

However, what is here is a little more than a straight reprinting. References to the Stata manuals have been updated. A few misprints have been corrected. ("Misprint" is British for "typo".) Notes following this *Preface* identify software changes since the columns were first produced. In any case, readers interested in identifying the most recent versions of any user-written programs should first use `search, all` within Stata.

Any writer accumulates many debts that cannot all be discharged or even acknowledged. William (Bill) Gould is the person who conceived of Stata and set up the company that maintains and develops it. StataCorp is a company that aspires to excellence in everything. As I discovered long ago, it is grateful even to be told of all bugs and misfeatures, down to misprints in its manuals. I remain perpetually aware that I follow in Bill's footsteps and stand on his shoulders.

I have also received much patient and detailed support from many other people at StataCorp. Vince Wiggins has borne the brunt of most of my specific graphics questions over the last decade, alerted me to some subtle tricks, and made encouraging noises throughout. Pat Branton, Lisa Gilmore, and their past and present colleagues have accommodated most of the quirks and dilatory submissions of an author with some strong prejudices about the English (N.B. not American) language and writing generally. Many in the Stata user community have asked good questions and provided most of the answers through encounters at users' meetings, postings on Statalist, and publications in the *Stata Technical Bulletin* and the *Stata Journal*. My fellow Editor Joe Newton has proved a master of the timely quiet word of approbation. Various specific acknowledgments are recorded in individual columns, but they form a sadly incomplete list. Although it is, as always, invidious to single out a few names, those who know anything of their work would want to echo further special thanks to Kit Baum, Marcello Pagano, and Patrick Royston for their long-standing endeavors for the Stata user community.

At Durham University, several geographical colleagues have kept me honest by providing a series of statistical challenges within a fascinating variety of datasets. Ian Evans in particular has also been an appreciative Stata user for many years. The division of labor in our collaborations that he did most of the more challenging science and I did most of the more challenging Stata programming has proved mutually congenial. It has been a special pleasure to remain in collaboration with my own PhD supervisor for over 40 years.

Most personally, I dedicate this book to my wife, Irene, in recognition of all the encouragement and support she has provided over our time together.

Nicholas J. Cox

Note: What follows is a complete list of my publications on Stata that are mostly graphical. As of this writing, all material published more than three years ago is freely available at http://www.stata-journal.com in the case of the *Stata Journal* and at http://www.stata.com/bookstore/individual-stata-technical-bulletin-issues/ in the case of the *Stata Technical Bulletin*. Graphical tips by me have been collected within Cox and Newton (2014).

SSG indicates reprinting in this book, and 119 indicates reprinting in Cox and Newton (2014).

Cox, N. J. 1997. gr16.1: Convex hull plots. *Stata Technical Bulletin* 36: 2–3. Reprinted in *Stata Technical Bulletin Reprints*, vol. 6, pp. 25–27. College Station, TX: Stata Press.

———. 1997. gr22: Binomial smoothing plot. *Stata Technical Bulletin* 35: 7–9. Reprinted in *Stata Technical Bulletin Reprints*, vol. 6, pp. 36–38. College Station, TX: Stata Press.

———. 1997. gr24: Easier bar charts. *Stata Technical Bulletin* 36: 4–8. Reprinted in *Stata Technical Bulletin Reprints*, vol. 6, pp. 44–50. College Station, TX: Stata Press.

———. 1997. gr24.1. Easier bar charts: correction. *Stata Technical Bulletin* 40: 12. Reprinted in *Stata Technical Bulletin Reprints*, vol. 7, p. 58. College Station, TX: Stata Press.

———. 1997. gr26: Bin smoothing and summary on scatter plots. *Stata Technical Bulletin* 37: 9–12. Reprinted in *Stata Technical Bulletin Reprints*, vol. 7, pp. 59–63. College Station, TX: Stata Press.

———. 1998. sg85: Moving summaries. *Stata Technical Bulletin* 44: 15–18. Reprinted in *Stata Technical Bulletin Reprints*, vol. 8, pp. 145–149. College Station, TX: Stata Press.

———. 1999. gr35: Diagnostic plots for assessing Singh-Maddala and Dagum distributions fitted by MLE. *Stata Technical Bulletin* 48: 2–4. Reprinted in *Stata Technical Bulletin Reprints*, vol. 8, pp. 72–74. College Station, TX: Stata Press.

———. 1999. gr41: Distribution function plots. *Stata Technical Bulletin* 51: 12–16. Reprinted in *Stata Technical Bulletin Reprints*, vol. 9, pp. 108–112. College Station, TX: Stata Press.

———. 1999. gr42: Quantile plots, generalized. *Stata Technical Bulletin* 51: 16–18. Reprinted in *Stata Technical Bulletin Reprints*, vol. 9, pp. 113–116. College Station, TX: Stata Press.

———. 2001. gr42.1: Quantile plots, generalized: update to Stata 7.0. *Stata Technical Bulletin* 61: 10–11. Reprinted in *Stata Technical Bulletin Reprints*, vol. 10, pp. 55–56. College Station, TX: Stata Press.

———. 2003. Software update: gr41_1: Distribution function plots. *Stata Journal* 3: 211.

——. 2003. Software update: gr41_2: Distribution function plots. *Stata Journal* 3: 449.

——. 2004. Software update: gr22_1: Binomial smoothing plot. *Stata Journal* 4: 490–491.

——. 2004. Software update: gr42_2: Quantile plots, generalized. *Stata Journal* 4: 97.

——. 2004. Speaking Stata: Graphing agreement and disagreement. *Stata Journal* 4: 329–349 (SSG).

——. 2004. Speaking Stata: Graphing categorical and compositional data. *Stata Journal* 4: 190–215 (SSG).

——. 2004. Speaking Stata: Graphing distributions. *Stata Journal* 4: 66–88 (SSG).

——. 2004. Speaking Stata: Graphing model diagnostics. *Stata Journal* 4: 449–475 (SSG).

——. 2004. Stata tip 6: Inserting awkward characters in the plot. *Stata Journal* 4: 95–96 (119).

——. 2004. Stata tip 12: Tuning the plot region aspect ratio. *Stata Journal* 4: 357–358 (119).

——. 2004. Stata tip 15: Function graphs on the fly. *Stata Journal* 4: 488–489 (119).

——. 2005. Software update: gr41_3: Distribution function plots. *Stata Journal* 5: 470–471.

——. 2005. Software update: gr42_3: Quantile plots, generalized. *Stata Journal* 5: 470–471.

——. 2005. Speaking Stata: Density probability plots. *Stata Journal* 5: 259–273 (SSG).

——. 2005. Speaking Stata: The protean quantile plot. *Stata Journal* 5: 442–460 (SSG).

——. 2005. Speaking Stata: Smoothing in various directions. *Stata Journal* 5: 574–593 (SSG).

——. 2005. Stata tip 21: The arrows of outrageous fortune. *Stata Journal* 5: 282–284 (119).

——. 2005. Stata tip 24: Axis labels on two or more levels. *Stata Journal* 5: 469 (119).

——. 2005. Stata tip 27: Classifying data points on scatter plots. *Stata Journal* 5: 604–606 (119).

——. 2006. Software update: gr26_1: Bin smoothing and summary on scatter plots. *Stata Journal* 6: 151.

——. 2006. Software update: gr42_4: Quantile plots, generalized. *Stata Journal* 6: 597.

——. 2006. Speaking Stata: Graphs for all seasons. *Stata Journal* 6: 397–419 (SSG).

——. 2007. Software update: gr0012_1: Density probability plots. *Stata Journal* 7: 593.

———. 2007. Speaking Stata: Turning over a new leaf. *Stata Journal* 7: 413–433 (SSG).

———. 2007. Stata tip 47: Quantile–quantile plots without programming. *Stata Journal* 7: 275–279 (119).

———. 2007. Stata tip 55: Better axis labeling for time points and time intervals. *Stata Journal* 7: 590–592 (119).

———. 2008. Speaking Stata: Between tables and graphs. *Stata Journal* 8: 269–289 (SSG).

———. 2008. Speaking Stata: Spineplots and their kin. *Stata Journal* 8: 105–121 (SSG).

———. 2008. Stata tip 59: Plotting on any transformed scale. *Stata Journal* 8: 142–145 (119).

———. 2009. Speaking Stata: Creating and varying box plots. *Stata Journal* 9: 478–496 (SSG).

———. 2009. Speaking Stata: Paired, parallel, or profile plots for changes, correlations, and other comparisons. *Stata Journal* 9: 621–639 (SSG).

———. 2009. Stata tip 76: Separating seasonal time series. *Stata Journal* 9: 321–326 (119).

———. 2009. Stata tip 78: Going gray gracefully: Highlighting subsets and downplaying substrates. *Stata Journal* 9: 499–503 (119).

———. 2009. Stata tip 82: Grounds for grids on graphs. *Stata Journal* 9: 648–651 (119).

———. 2010. Software update: gr0009_1: Speaking Stata: Graphing model diagnostics. *Stata Journal* 10: 164 (SSG).

———. 2010. Software update: gr0021_1: Speaking Stata: Smoothing in various directions. *Stata Journal* 10: 164 (SSG).

———. 2010. Software update: gr41_4: Distribution function plots. *Stata Journal* 10: 164.

———. 2010. Software update: gr42_5: Quantile plots, generalized. *Stata Journal* 10: 691–692.

———. 2010. Speaking Stata: Graphing subsets. *Stata Journal* 10: 670–681 (SSG).

———. 2010. Speaking Stata: The statsby strategy. *Stata Journal* 10: 143 151 (SSG).

———. 2011. Stata tip 102: Highlighting specific bars. *Stata Journal* 11: 474–477 (119).

———. 2011. Stata tip 104: Added text and title options. *Stata Journal* 11: 632–633 (119).

———. 2012. Software update: gr42_6: Quantile plots, generalized. *Stata Journal* 12: 167.

———. 2012. Speaking Stata: Axis practice, or what goes where on a graph. *Stata Journal* 12: 549–561 (SSG).

———. 2012. Speaking Stata: Transforming the time axis. *Stata Journal* 12: 332–41 (SSG).

———. 2013. Speaking Stata: Creating and varying box plots: Correction. *Stata Journal* 13: 398–400 (SSG).

———. 2013. Speaking Stata: Trimming to taste. *Stata Journal* 13: 640–666 (SSG).

———. 2014. Stata tip 119: Expanding datasets for graphical ends. *Stata Journal* 14: 230–235 (119).

Cox, N. J., and N. L. M. Barlow. 2008. Stata tip 62: Plotting on reversed scales. *Stata Journal* 8: 295–298 (119).

Cox, N. J., and A. R. Brady. 1997. gr25: Spike plots for histograms, rootograms, and time-series plots. *Stata Technical Bulletin* 36: 8–11. Reprinted in *Stata Technical Bulletin Reprints*, vol. 6, pp. 50–54. College Station, TX: Stata Press.

———. 1997. gr25.1: Spike plots for histograms, rootograms, and time-series plots: update. *Stata Technical Bulletin* 40: 12. Reprinted in *Stata Technical Bulletin Reprints*, vol. 7, p. 58. College Station, TX: Stata Press.

Cox, N. J., and H. J. Newton, eds. 2014. *One Hundred Nineteen Stata Tips*. Stata Press, College Station, TX.

Gray, J. P., and N. J. Cox. 1998. gr16.2: Corrections to condraw.ado. *Stata Technical Bulletin* 41: 4. Reprinted in *Stata Technical Bulletin Reprints*, vol. 7, p. 58. College Station, TX: Stata Press.

Royston, P., and N. J. Cox. 2005. A multivariable scatterplot smoother. *Stata Journal* 5: 405–412.

Sasieni, P., P. Royston, and N. J. Cox. 2005. Software update: sed9_2: Symmetric nearest neighbor linear smoothers. *Stata Journal* 5: 285.

Software notes on the original columns

This page collects some notes on important changes to commands since the columns were originally published. These notes were written in March 2014.

Official commands

The `levels` command is mentioned in article 1, "Speaking Stata: Graphing distributions". `levels` still works but is undocumented. It has been superseded by `levelsof`.

Biplots are mentioned in article 2, "Speaking Stata: Graphing categorical and compositional data". Since that was written, Stata has added support for various kinds of biplots. `search biplot` in Stata will get you started.

The user-written command `locpoly` is mentioned in article 4, "Speaking Stata: Graphing model diagnostics". It is now superseded by the official command `lpoly`.

User-written commands by the author

The command `onewayplot` is mentioned in article 1, "Speaking Stata: Graphing distributions". It is now superseded by `stripplot`. Both commands can be downloaded from Statistical Software Components (SSC) using `ssc install onewayplot` or `ssc install stripplot`.

The command `catplot` is mentioned in article 2, "Speaking Stata: Graphing categorical and compositional data", and article 10, "Speaking Stata: Spineplots and their kin". It can be downloaded from SSC using the command `ssc install catplot`. `catplot` as rewritten in 2010 has a different syntax from that used in examples in those articles.

The command `distplot` is mentioned in article 2, "Speaking Stata: Graphing categorical and compositional data". It can be downloaded from the *Stata Journal* archive following `search distplot`; choose the most recent update. `distplot` as rewritten in 2005 has a different syntax from that used in examples in that article.

The command `tableplot` is mentioned in article 2, "Speaking Stata: Graphing categorical and compositional data". It is now superseded by an extended `tabplot`. Both commands can be installed from SSC using the `ssc` command, as mentioned above.

The Stata Journal (2004)
4, Number 1, pp. 66–88

Speaking Stata: Graphing distributions

Nicholas J. Cox
University of Durham, UK
n.j.cox@durham.ac.uk

Abstract. Graphing univariate distributions is central to both statistical graphics, in general, and Stata's graphics, in particular. Now that Stata 8 is out, a review of official and user-written commands is timely. The emphasis here is on going beyond what is obviously and readily available, with pointers to minor and major trickery and various user-written commands. For plotting histogram-like displays, kernel-density estimates and plots based on distribution functions or quantile functions, a large variety of choices is now available to the researcher.

Keywords: gr0003, graphics, histogram, spikeplot, dotplot, onewayplot, kdensity, distplot, qplot, skewplot, bin width, rug, density function, kernel estimation, transformations, logarithmic scale, root scale, intensity function, distribution function, quantile function, skewness

1 Introduction

The new graphics introduced in Stata 8 has been, by far, the most important step forward in Stata's graphical functionality since early releases in the mid-1980s. It is, therefore, high time that this column turned to discuss graphics directly. I intend to make 2004 a graphic year for *Speaking Stata*, starting with the basic and fundamental issue of graphing univariate distributions. Future columns are intended to discuss graphing categorical and compositional data, comparisons, and model diagnostics. In each case, the aim will be to provide an overview of Stata's provision and to show ways to go beyond what is obviously and readily available. The emphasis will be on graphics commands of potential interest to the largest possible cross-section of Stata users. Thus histograms clearly qualify, but justice cannot be done to details specific to analysis of survival-time distributions.

The core commands for graphing distributions range from `twoway kdensity` and its relative `kdensity` through `twoway histogram` and its relative `histogram` to `graph box` and `graph hbox`. Related but perhaps less-often used commands include `dotplot`, `spikeplot`, and those grouped as diagnostic plots.

2 Histograms, indigenous and exotic

2.1 Number of bins and bin width

With an eye to tradition, including Stata tradition, let us start the discussion with histograms. Up until Stata 7, a histogram was the default graph type if `graph` was fed just one variable. Before Stata 8, such histograms were relatively inflexible and could not easily be combined with other graph types. Now we have both greater flexibility and easier working with other types. Notable additions include the options to tune both bin width and the start of binning, whereas previously only the number of bins could be controlled directly. The start of binning could be controlled indirectly by tuning `xlabel()` or `xscale()`.

As every good introductory text explains, histogram construction is largely a trade-off problem in which you seek a compromise between detail and generalization or between variance and bias. In doing this, you can tune either the number of bins or the bin width. Theoretical discussions concentrate on the number of bins and its relation to sample size and the kind of distribution being analyzed. However, my guess is that people with their feet in application areas often find it natural to think in terms of a sensible bin width for the variables they have, bearing in mind measurement issues and the magnitude of important or interpretable differences. Whatever your preference, you can now do it either way.

2.2   Varying bin widths

However, one feature that remains wired in histogram commands in Stata 8 is a restriction to bins of equal width. No doubt this is often very sensible whenever the original data are available, but there are occasions on which you might want to break this rule. Let us drill down to some first principles here.

Recall that the idea behind histograms is that the area of each bar represents the fraction of a frequency (probability) distribution within each bin (or class, or interval). Among many books explaining histograms, Freedman, Pisani, and Purves (1978) is an outstanding introductory text that strongly emphasizes the area principle. It is not part of the definition that all bins have the same width, but rather that what is shown on the vertical axis is, or is proportional to, probability density. Frequency density qualifies, as does frequency if all bins have the same width.

In practice, the choice of bin width is often a little arbitrary. If the variable is discrete, a width of 1 is clearly a natural choice. Even then, discrete variables may require some grouping into bins wider than 1. If the variable is number of lifetime sexual partners, the tail (apparently) stretches into very large numbers, and some grouping may be desired. With continuous variables especially, there is always some arbitrariness. Many researchers are most reluctant to compound that by varying the width of the intervals. To do so would complicate the interpretation of the histogram, it might be argued, by any variations in the way the bars were produced. Or, to put it another way, equal widths are relatively simple, and any kind of complexity beyond them needs to be justified.

Despite all that, sometimes the data come grouped into irregular intervals, and the researcher has little or no choice because the raw data may be difficult or impossible to access. Sometimes there is an underlying confidentiality issue. Nevertheless, researchers may still want a histogram, which should be correctly drawn with density, not frequency, on the vertical axis. For example, Altman (1991, 25) gives the ages of 815 road accident casualties for the London Borough of Harrow in 1985:

age	frequency
0–4	28
5–9	46
10–15	58
16	20
17	31
18–19	64
20–24	149
25–59	316
60+	103

In this example and in other similar examples, density can only be calculated for the open-ended class if we specify an upper limit; Altman suggests that 60+ be treated as 60–80.

As usual in statistics, sampling variation is also an issue. If we regard the histogram as a crude estimator of a density function, there is often a case for varying bin width to match the structure of variation, in effect varying how we average probability density locally.

But there is at least one other way to build a histogram in a simple, systematic way: using as limits a set of quantiles equally spaced on a probability scale (e.g., Breiman 1973, 208–209; Scott 1992, 69–70). That way, each bar represents the same area. Unless our data come from something like a uniform distribution, the bin widths will be markedly unequal, but they will reflect the character of the distribution. Breiman points out that the associated error will be approximately a constant multiple of the bar heights, so long as the bin frequencies are not too small.

A related problem is choice of class intervals for a chi-squared test of goodness of fit. Mann and Wald (1942) and Gumbel (1943) urged the merits of choosing classes with equal expected frequencies. That is a simple and definite procedure, which can reduce difficulties arising from low expected frequencies, although data must arrive ungrouped and there may be some loss of sensitivity in the tails of a distribution. Without getting into a wider discussion of the merits of different tests of fit or of tests compared with graphical analysis, it is clear that the equal probability idea is a natural one.

What can be done in Stata? Start with the messier problem in which the data arrive grouped. Much can be done once you know about an undocumented feature of `twoway bar`. We need to enter the lower bin limits and the bin frequencies and one final upper limit as data. For Altman's example, we need to enter data to get

```
. list age freq
```

	age	freq
1.	0	28
2.	5	46
3.	10	58
4.	16	20
5.	17	31
6.	18	64
7.	20	149
8.	25	316
9.	60	103
10.	80	.

We then can calculate the densities:

```
. generate density = freq / (815 * (age[_n+1] - age))
```

If you want frequency density rather than probability density, you should omit scaling by the sample size (here 815).

Finally, we can draw the graph, shown in figure 1:

```
. twoway bar density age, bartype(spanning) bstyle(histogram)
```

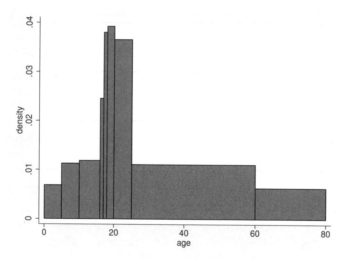

Figure 1: Example of histogram based on data supplied as frequency distribution with varying bin widths.

The "spanning" extends bars to the right until they are curtailed; that is why it is necessary to specify all lower limits and one upper limit for the graph. The data should also be in the correct sort order, as in this example. The option `bstyle(histogram)` is not compulsory, and you might like to check other possibilities. You might need to add the option `yscale(range(0))` if `twoway bar` does not automatically start bars at 0.

Turning to the more elegant problem, a user-written program for equal-probability histograms can be described and, if desired, downloaded from the Statistical Software Components (SSC) archive by using the `ssc` command; see [R] **ssc**:

```
. ssc describe eqprhistogram
. ssc install eqprhistogram
```

As an illustration, figure 2 is the result of

```
. use http://www.stata-press.com/data/r8/womenwage.dta
. eqprhistogram wage, bin(10) plot(kdensity wage, biweight width(5))
>     legend(ring(0) position(1) column(1))
```

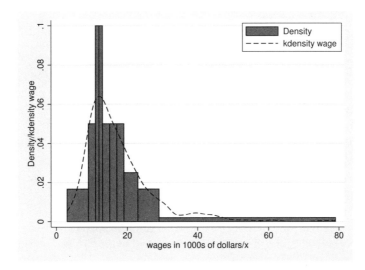

Figure 2: Example of histogram in which bins represent intervals of equal probability. In this case, bin boundaries are deciles, so that the area of each bin represents 1/10 of the distribution. A kernel-density estimate is superimposed.

The bin limits are the deciles, so each bar represents 1/10 of the total probability in the distribution. Note that you can superimpose a density estimate.

Although it may seem a curiosity, the equal-probability histogram has some pedagogic merit. First, it underlines the area principle on which histograms are based. Wider bars are necessarily shorter and narrow bars necessarily taller. Second, it allows a link to be made between histograms and quantile-based methods such as box plots. Arguably, in some datasets it gives a better view of the tails than do the corresponding box plots, especially if within those box plots no values are flagged beyond the quartiles, and so no details are given on structure within the tails. (Box plots are especially poor for U-shaped distributions. In some such cases, no values are identified beyond the quartiles, and the box plot reduces to a long box and two short tails. Even experienced people can misread this as indicating a unimodal distribution, forgetting that if half the values lie inside the box, then, necessarily, half lie outside it.)

An equal-probability histogram is not suitable for all distributions. Given categorical, discrete, or highly rounded data, quantiles may be tied, especially if the number of bins is large relative to the sample size. If the specified quantiles are tied, `eqprhistogram` refuses to draw the graph. A technical aside: whenever it does this, the exit code is 0. This is in part a diplomatic acknowledgment that inability to draw the graph is either a feature of the data or a limitation of the method, rather than a user error. In addition, it implies that a loop through equal-probability histograms of different variables or groups will not fail merely because a particular graph is impossible.

2.3   Putting a rug underneath

One major merit of histograms is familiarity. All statistically minded people have looked at many histograms, and nonstatistical people who use statistics have also usually come across them. Nevertheless, the basis of histograms, a division of a range into bins, is at best a means to an end, namely easy and effective visualization of a distribution, and at worst a serious distraction. Both psychologically and numerically, densities or frequencies calculated from a set of bins can convey a poor idea of the detailed shape of the distribution of a variable.

One simple way to enhance a histogram by forging a closer link with the raw data is to add a so-called rug, which as the name implies, is almost always placed underneath the histogram. A rug is a very short, long display of point symbols, one for each distinct value. Often a vertical pipe symbol | is used to minimize overlap. Rugs may also be added to other kinds of plots. There are many varied examples in Davison (2003).

Before version 8, Stata had `graph` options to combine rugs, which in Stata were called oneway plots, with box plots and with scatterplots. These options are still accessible under `graph7`. However, they did not make the cut into the new graphics in that or similar form.

Although rugs are not explicitly provided in Stata 8, the procedure for weaving your own rug is straightforward. Starting with a basic histogram for the same wage data,

```
. histogram wage, start(0) width(5)
```

we see that with these choices density varies up to about 0.07 per 1,000 dollars. That leads to a decision to put the rug at about −0.003 on that scale. We need a variable to hold this value:

```
. generate where = -0.003
```

In practice, we can just choose a trial value and then use `replace` to improve upon it. Next, there is no pipe symbol in the `symbolstyle` portfolio, so we must enlist the pipe character as a marker label. Then, the rug is just a scatterplot of `where` against `wage`, suppressing the default marker symbol and placing the marker label exactly on target:

```
. generate pipe = "|"
. histogram wage, start(0) width(5)
>     plot(scatter where wage, ms(none) mlabel(pipe) mlabpos(0)) legend(off)
```

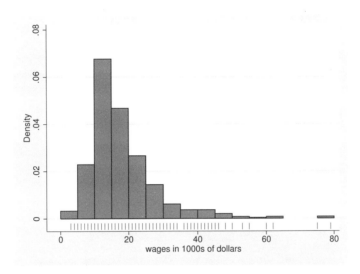

Figure 3: Example of histogram with rug showing distinct values occurring in the data.

In this case, as shown in figure 3, the rug shows rounding of the data, and a tabulation makes it explicit that all values are just multiples of $1,000. In general, a rug is a useful but restrained way of showing some of the fine structure of a distribution.

A rug will take up a lot of bytes in a graph file if any point symbol stands for many repeated values. Clearly, it is unnecessary to overwrite each symbol repeatedly. A solution is to select each distinct value just once. There are two systematic ways to do this, to select the first in each group after sorting or to select the last, and it is immaterial here which you use, so you might as well go

```
. bysort wage: generate tag = _n == 1
```

A canned near-equivalent is

```
. egen tag = tag(wage)
```

The difference is that the `egen` call sorts your data while doing the calculation but then returns it to its original sort order, which may differ. The first method may change your sort order. Having done this, we select points for the rug as `if tag`.

2.4   Horizontal histogram bars

The `histogram` display, by default, has the frequency axis vertical, as is conventional. The manual entry [G-2] **graph combine** shows how histograms may be placed vertically and horizontally along the margins of a scatterplot. More generally, the `horizontal` option may be used to reverse axes. This may sound merely cosmetic, but there are occasions in which this layout appears more natural. In the environmental sciences, among other fields, height above and depth below some surface are key natural variables. The extra option `yscale(reverse)` would show depths the intuitive way up.

Here is a histogram of the mean elevations of 27,523 glaciers from Central Asia and southern Siberia (figure 4). Data were extracted from the World Glacier Inventory. The tendency to multiple modes is best interpreted as a consequence of lumping together several distinct mountain ranges. The Stata command was

```
. histogram mean_elev, horizontal start(1600) width(100) frequency
>      ylabel(, angle(horizontal))
```

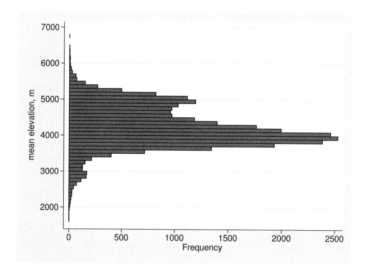

Figure 4: Example of histogram with horizontal bars. In this case, the response variable is altitude.

2.5   Do-it-yourself density calculation

From a simple enhancement of histograms, we turn to the basic underlying calculations. It may be useful to document how to calculate densities yourself, having first chosen a *start* and a *width*. If we are counting so that each bin is defined precisely as *lower_limit* $\leq$ *value* < *upper_limit*, we could use `floor()` to generate lower limits as bin identifiers. With the reverse convention, we could use `ceil()`. See Cox (2003c) for a note on these functions. Then, the frequencies are the counts within each bin:

```
. generate double lower = width * floor((varname - start) / width)
. bysort lower: generate frequency = _N
```

To get fractions and percents, we must be careful to count each value just once:

```
. by lower: generate double sum = frequency * (_n == 1)
. replace sum = sum(sum)
. generate fraction = frequency / sum[_N]
. generate percent = 100 * fraction
```

The density is then

```
. generate density = frequency / (width * _N)
```

Real calculations get messier once you build in selections if or in, subdivision into groups defined by other variables, or missing values. The messy details are coded up in an egen function density() in the egenmore package on SSC.

3 Relatives of histogram: spikeplot, dotplot, and onewayplot

One common reaction to histograms is to prefer more information in displaying distributions, especially in the tails. The optimistic view is that more details will turn out to be instructive fine structure. The corresponding pessimistic view is, naturally, that such details will be best regarded as noise and, as such, an irreducible nuisance. Most discussions stress the latter view over the former, but there can be real merit in playing deterministic detective rather than stochastic skeptic.

The official commands spikeplot and dotplot and the user-written command onewayplot offer different ways of showing more detail than do equivalent histograms.

spikeplot, by default, offers a spike for every distinct value—that is, no binning—and the opportunity to control binning by a round() option, which in effect controls bin width. Historically, spikeplot offered, before Stata 8, the most obvious official alternative to graph, histogram for getting a histogram-like display with more than 50 bins. That role is now lost. However, its discrete representation of a frequency distribution remains available for occasions when you want to emphasize the granularity of data, either as defined in principle (counted variables, in particular) or as measured in practice. The display of the age distribution of Ghana given at [R] **spikeplot** is a good example of what spikeplot does best, revealing a fine structure of age preferences, including multiples of 5, even rather than odd ages, and so forth. There is some scope for controlling spike appearance if the default appearance (which is the default of twoway spike under the prevailing scheme) appears too exiguous.

dotplot, in contrast, is based on the idea (or the ideal) of showing a point symbol for each value; exactly the same description covers rugs and onewayplot. Similar plots under a variety of names go back at least as far as van Langren (1644); see Tufte (1997, 15). Wilkinson (1999) gives several further references of historical interest. Chambers et al.

(1983) used the term one-dimensional scatterplots. The term oneway plots appears to have been introduced by StataCorp in its earlier guise as Computing Resource Center (1985). Wild and Seber (2000) show many interesting examples of oneway plots.

dotplot, by default, offers, as far as possible, a point symbol for every value and some binning. Binning can be controlled rather indirectly, although in practice, the default is usually adequate, and when desired, the binning can be switched off with the nogroup option. The main virtues of dotplots lie in their ability to show some features that might otherwise be obscured by a series of touching bars, especially granularity and details of outliers or other extreme values in the tails. You can also show, for example, median and quartiles by horizontal marks and thus hybridize box plots and histograms.

The considerable flexibility of histogram, spikeplot, and dotplot might seem to leave few important gaps in their territory. Nevertheless, onewayplot was written to provide some extra possibilities in this area; it also may be downloaded using ssc. As mentioned earlier, graph, oneway did not survive as such into Stata 8, although the minor trickery needed to add rugs is just one illustration of how they can be emulated fairly easily. onewayplot is essentially a convenience command that bundles together various easy but tedious handles for making your own oneway plots. You can choose between horizontal and vertical layouts, while stack and center options produce a variant on dotplot.[1] There is, by default, no binning of data; binning may be accomplished with the width() option.

In figure 5, we show the results of a onewayplot using the handle of a regional classification to split the glacier elevations. Both histogram and dotplot struggle given 18 regions, some with fairly long names.

```
. onewayplot mean_elev, by(region) ytitle("") stack ms(oh) msize(tiny) width(20)
```

---

1. You can also type centre. An undocumented feature of dotplot is that centre is allowed as well as center. This is a convenience for speakers of languages, such as English, which use that spelling, and is emulated by onewayplot.

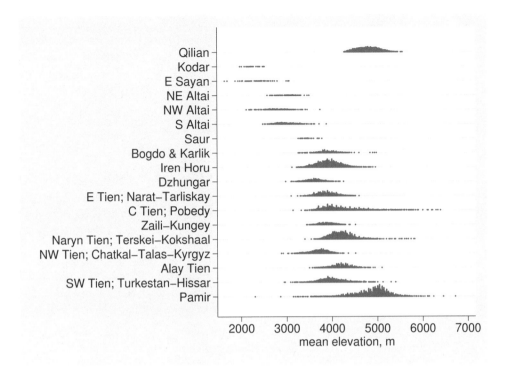

Figure 5: Example of multiple histograms produced by onewayplot.

4 Kernel-density estimation

4.1 Available commands

The histogram of a continuous variable is, from one point of view, an estimator of the density function of that variable. Clearly the set of bins used to compute that estimate imposes discontinuities on the estimate, which leads us directly to consider smoother estimates, especially those based on convolution of the data and a symmetric kernel. twoway kdensity and kdensity are provided in official Stata as basic commands. Recently, users have added variable kernel density estimation commands (Salgado-Ugarte and Pérez-Hernández 2003; Van Kerm 2003).

4.2 Variations on the official theme

Transform before and after estimation

Some simple devices extend the range of applications of Stata's official commands for kernel-density estimation. First is the idea of estimating the density function on a transformed scale and then back-transforming the estimate to one for the raw scale. Two of the most natural transformations here, as elsewhere, are logarithms for positive variables and logit-like transformations for proportions and other data measured on some interval (a, b). The underlying general principle is that, for a continuous monotone transformation $t(x)$, the densities $f(x)$ and $f\{t(x)\}$ are related by $f(x) = f\{t(x)\}|dt/dx|$. This procedure is mentioned briefly by Silverman (1986, 27–30), although his worked example (page 28) is not very encouraging. Good expositions are given by Wand and Jones (1995, 43–45), Simonoff (1996, 61–64), and Bowman and Azzalini (1997, 14–16).

With a logarithmic transformation of x, we have

$$\text{estimate of } f(x) = \text{estimate of } f(\log x) \times (1/x)$$

given that $d/dx(\log x) = 1/x$. Note in particular, if data are right skewed, that the result of this transformation is more smoothing in the tail and less near the main part of the distribution than in the default method. I have found this to be one of the most valuable ways of going beyond the default. It fits very well both the common finding that positive variables are right-skewed, suggesting a transformation, such as the logarithm, and the common attitude that results on the original scale are of direct scientific or practical interest. To put it another way, the transformation behaves more like a link function than a classical transformation, given that end results are on the scale of the original response. You can get the best of both worlds.

Returning to the wage data, here is an illustrative (and certainly not definitive) example, in which we just use default kernel and width choice.

```
. generate logwage = log(wage)
. kdensity logwage, at(logwage) generate(densitylog)
. generate density = densitylog/wage
. levels wage, local(levels)
. line density wage, sort xtick(`levels´, tposition(inside))
```

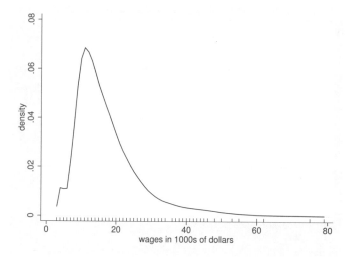

Figure 6: Example of a density function estimated on logarithmic scale and transformed to the original scale of data.

The density function, shown in figure 6, is much smoother in the tails than the equivalent default, which is not shown here. However, the step in the left-hand tail needs investigation: is this some odd artifact or a genuine feature of the data? Incidentally, another technique is used to show a rug by picking up a list of distinct values from `levels` (added to Stata on 16 April 2003). However, this technique is not as general as that previously illustrated, as it hinges on the variable concerned having only integer values. `levels` is not designed to work with noninteger numeric values.

Similarly with a logit-like transformation,

$$\text{estimate of } f(x) = \{\text{estimate of } f(\text{logit } x)\} \, \frac{(b-a)}{(x-a)(b-x)}$$

where logit $x = \log\{(x-a)/(b-x)\}$, a slight generalization of the usual definition, for which $a = 0, b = 1$. Note that $d/dx(\text{logit } x) = (b-a)/\{(x-a)(b-x)\}$.

Density on a log scale

It can be natural to calculate $f(\log x)$ as a way of getting a better estimate of $f(x)$. It can also be natural to calculate $\log\{f(x)\}$ as way of getting a better visualization of $f(x)$. This seems to be an old idea, periodically rediscovered. One venerable reference is the work of the soldier, explorer, and scientist R. A. Bagnold (1937, 1941), who worked on size distributions—especially those of sand—while at present the idea is widely used in fields ranging from statistical physics (Bardou et al. 2002) to statistical finance (Hazelton 2003). There is clearly no barrier also to looking at $\log\{f(\log x)\}$ or using some other transformation before density estimation if it seems appropriate.

The highly original contribution of Bagnold deserves some explanation, as it appears to be little known within the statistical sciences. Born in England in 1896[2], he joined the British army from school and served in the First World War. He then took an engineering degree at Cambridge. Remaining in the army, he used leaves to travel and explore, particularly on pioneering long trips into the deserts of Egypt, Sudan, and Libya using specially adapted cars. This provoked an interest in the physics of blown sand, leading ultimately to a now-classic monograph (Bagnold 1941). Wind transport of loose particles is highly size selective, as ordinary experience confirms: very coarse material will not move, while very fine material may easily be lofted high into the atmosphere and carried over vast distances. Thus, the particle size distribution of a deposit (say from a sand dune) is of central interest. Bagnold found plots of log density versus log grain diameter the most helpful way to show his data. It seems clear from his very readable autobiography (1990), published just after his death, that the crucial first step of plotting densities on a log scale owed most to an engineer's feeling of a sensible thing to do. By thinking for himself, he was not inhibited by ideas on what was or was not standard statistical practice. Much later, Bagnold returned to the question and contributed to the development of log-hyperbolic distributions by Ole Barndorff-Nielsen. There is a full bibliography of his publications in Thorne et al. (1988).

---

2. His sister was Enid Bagnold, later a novelist, dramatist, and poet, and best remembered for the children's classic *National Velvet*.

Several properties are simple on a log-density scale. One exploited by Bagnold is that a normal (Gaussian) density plots as a parabola

$$\log f(x) = -\log\left(\sigma\sqrt{2\pi}\right) - \frac{(x - \mu)^2}{2\sigma^2}$$

while exact or approximate exponential or power-law decay of density will show exact or approximate linear patterns, the latter requiring also a logarithmic scale for the variable.

Let us illustrate with log of wage from the wage data considered above:

```
. generate logdensitylog = log(densitylog)
. quietly summarize logwage
. local mean = r(mean)
. local sd = r(sd)
. scatter logdensitylog logwage
>           || function normal =
>           -log(`sd´ * sqrt(2 * _pi)) - ((x - (`mean´))^2 / (2 * `sd´^2)),
>           ra(logwage) ytitle(log density) xtitle(log wage)
>           legend(off) subtitle(log density plot)
```

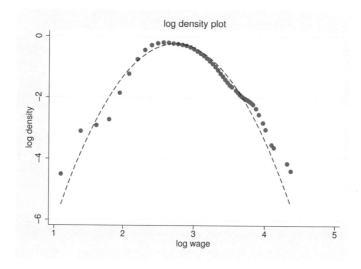

Figure 7: Example of plot using log density. The parabola shows a normal density function with the same mean and standard deviation as log wage.

The results in figure 7 suggest a good but not spectacularly good fit to a lognormal. The slightly fat tails seem suggestive. At the same time, the density estimates, especially in the tails, are, as always, subject to sampling variation and sensitive to kernel bandwidth; note also that neighboring density estimates are necessarily highly dependent.

What is implemented above is just a first stab. Hazelton (2003) suggests various refinements, including robust estimation of the mean and standard deviation and, given a density estimation bandwidth h, fitting a normal with variance $\mathrm{sd}^2 + h^2$ to correct for side-effects of using a kernel.

Density on a root scale

There would also be some advantages to a square-root scale, given that densities behave a bit like counts, for which a root transformation is often the first to be tried. Also, the square root of a Gaussian shape is another Gaussian shape. So we can have our cake and eat it too: hunt for a Gaussian yet benefit from stabilized sampling fluctuation. Check the assertion with

```
. twoway function sqrt(normden(x)), range(-4 4)
```

There is a `root` option in `spikeplot` for a similar reason. Tukey (1977, chapter 17) worked through a bundle of related ideas, which seem to have been little explored since.

Intensities, too

Those interested in data on events, considered as the result of a point process in one dimension (most obviously, time or space), should note that Stata's kernel-density commands can readily be used to estimate the intensity function (say, frequency per unit of time or space). Suppose that a variable contains dates of earthquakes, eruptions, strikes, honors for a sports team, or whatever else is of interest. To get results on an intensity scale, just multiply 'density' by the number of observed data points. A key detail is that intensities will be smoothed beyond the beginning and the end of the interval in question; whether this is tolerable or further surgery is desired is a question for the user.

5   Quantile plots and distribution plots

Another key approach eschews any kind of binning or smoothing and starts with the idea of directly showing the pattern of the quantiles (the ordered values). Formally, we order n data values for a variable x and label them such that $x_{(1)} \leq \cdots \leq x_{(n)}$. `quantile` has long been available as an official Stata command for quantile plots, in which the $x_{(i)}$ are plotted against $(i - 0.5)/n$. See [R] **diagnostic plots**. The term *quantile plot* appears in Chambers et al. (1983) and Cleveland (1993, 1994). Modern use of quantile plots and their relatives stems largely from the path-breaking paper of Wilk and Gnanadesikan (1968). Examples of antecedents from the nineteenth century can be found in Quetelet (1827) and Galton (1875); see Stigler (1986, 167, 270).

Essentially the same information can be shown in a plot of cumulative distribution functions or of survival (a.k.a., survivor, reliability, complementary, or reverse distribution) functions, in which we plot either probabilities or frequencies of values being $\leq x$

or $> x$. In many biomedical or engineering applications, the survival function appears closer to the practical problem, the name being suitably evocative if data are indeed times to patient death, component failure, or something similar.

According to Hald (1990, 108), the first graph of (the complement of) a distribution function appears in a 1669 letter from Christiaan Huygens (1629–1695) to his brother Lodewijk (1631–1699). He plotted a survival function from data from the life table of John Graunt (1620–1674). Huygens made numerous contributions to mathematics, astronomy, and physics, studying, among many other matters, games of chance, the collision of elastic bodies, the rings of Saturn, the pendulum clock, and the wave theory of light.

In Stata 8, the graphics of `quantile` were revised to match the new graphics, but the functionality was unchanged. A broader command is `qplot` (Cox 2004), which in most respects is a generalization of `quantile`; just one detail is omitted, the reference line. It supersedes the previous program `quantil2` (Cox 1999b, 2001).

Stata already has an official graph command for survival functions, `sts graph`. If your data really are survival times and you have any of the complications that are the stuff of survival analysis, such as censoring or subjects entering at different times, you should use `sts graph`. However, it is not and does not purport to be a general purpose command for all kinds of distribution.

In addition, the official command `cumul` ([R] **cumul**) is available to calculate the cumulative distribution function for a single variable, after which the function may be plotted using `twoway`. The user with several variables to be compared or with an interest in survival functions thus needs to repeat the `cumul` command or take the further step of calculating survival functions from cumulative distribution functions. A command `distplot` that bundles calculation and graphing steps together is, however, available (Cox 1999a, 2003a,b).

`qplot` and `distplot` are, in effect, siblings. The choice between them is most obviously one of choice of axes and thus, in a sense, trivial, but different conventions may seem natural for different problems and even different fields or traditions. In particular, there seems to be a growth of interest in quantile functions as responses, which makes `qplot` a possible choice (see, for example, Gilchrist 2000).

`qplot` and `distplot` have in common

1. Support for graphing several variables.

2. Support for graphing several groups, through a `by()` option.

3. Choice of `twoway` plottypes, from `area`, `bar`, `connected`, `dot`, `dropline`, `line`, `scatter`, or `spike`. These are not in general equally useful or attractive, but there is at least much choice, courtesy of `twoway`'s generous design.

4. Support for reversing the sort order so that values decrease from top left.

5. Support for alternative transformed scales.

In addition, `qplot` has support for choice of a in a general rule for plotting position $(i - a)/(n - 2a + 1)$ for $i = 1, \ldots, n$. The default is $a = 0.5$, giving $(i - 0.5)/n$. Other choices include $a = 0$, giving $i/(n + 1)$, and $a = 1/3$, giving $(i - 1/3)/(n + 1/3)$. The choice is often immaterial, but some authorities have strong opinions on the best choice on various grounds, some even statistical. For more discussion and references, see Cox (1999b).

6   Skewness plots

The skewness of a variable is often of interest, perhaps especially as an indicator of potential problems in subsequent analysis. Commonly a single measure is used, whether the moment-based measure produced by `summarize, detail` or other measures (which, in most cases, are readily calculated from the output of `summarize`). Graphically, skewness may be assessed with varying degrees of ease and efficiency from the plots mentioned so far, but there is also a case for a customized design.

Various possibilities are based on the quantiles (Gnanadesikan 1977, 1997). The quantiles may be paired as lower and upper quantiles $x_{(1)}$ and $x_{(n)}$, $x_{(2)}$ and $x_{(n-1)}$, etc., and a median may be calculated in the usual way.

Stata supports `symplot`, a plot of (upper quantile − median) versus (median − lower quantile), for which the reference situation of symmetry or lack of skew plots as a line of equality. See [R] **diagnostic plots**. However, `symplot` will show only a single group of data and thus cannot be used for comparisons, while a plot with a sloping reference line is more difficult to deal with than the plot now to be described, which has horizontal reference lines.

`skewplot` produces, by default, a plot of the *midsummary* versus the *spread* for the variables supplied, also known as the mid-versus-spread plot. With the `skew` option, it produces a plot of the *skewness function* versus the *spread function*. Such plots convey both the general character and the fine structure of the symmetry or skewness of datasets and can be used to compare distributions or to assess whether transformations are necessary or effective.

There are some little-used terms here, so we need a few definitions. In a perfectly symmetric set of data, the midsummaries $(x_{(1)} + x_{(n)})/2$, $(x_{(2)} + x_{(n-1)})/2$, etc., would all be identical and equal to the median. A plot of each midsummary (or mean of lower and upper quantiles) $(x_{(i)} + x_{(n-i+1)})/2$ versus each difference or spread of lower and upper quantiles $x_{(n-i+1)} - x_{(i)}$ would, thus, yield a horizontal straight line. Conversely, skewness in sets of data will be reflected by departures from horizontality. In particular, right skewness would be shown by rising lines and left skewness by falling lines.

Apart from the divisor of 2, this plot was suggested by J. W. Tukey (Wilk and Gnanadesikan 1968). See also Gnanadesikan (1977, 1997, chapter 6.2) or Fisher (1983). The form used here and the name *mid-versus-spread plot* are found in Hoaglin (1985). It is usual to plot only that half of the sample results for which spread is ≥ 0.

The `skew` option produces an alternative form promoted by Benjamini and Krieger (1996, 1999). Consider the identity, which introduces their terminology,

$$
\begin{aligned}
x_{(n-i+1)} &= \text{median} + \left(x_{(n-i+1)} - x_{(i)}\right)/2 + \left(x_{(i)} + x_{(n-i+1)} - 2 \times \text{median}\right)/2 \\
&= \text{median} + spread\ function + skewness\ function
\end{aligned}
$$

for $x_{(i)}$ in the lower half of a sample. This leads to a plot of the skewness function versus the spread function, known as the skewness versus spread plot. Note that the skewness function is (midsummary − median) and so will be constant and zero for a perfectly symmetric distribution and that the spread function is half the spread of the mid versus spread plot. In short, the `skew` option does not change the configuration of the plot but merely the labeling of the axes.

In addition, the ratio of the skewness and spread functions or

$$
\frac{x_{(i)} + x_{(n-i+1)} - 2 \times \text{median}}{x_{(n-i+1)} - x_{(i)}}
$$

is a measure of skewness (in the traditional sense) originally suggested for quartiles by Bowley (1902) and generalized to this form by David and Johnson (1956). Another incarnation is as the p-skewness index (Gilchrist 2000, 54, 72).[3] It varies between -1 and 1. A similar general measure was used by Parzen (1979). Graphically this measure is the slope of the line connecting $(0,0)$ and each data point if the `skew` option is used.

See Benjamini and Krieger (1996, 1999) and Groeneveld (1998) for concise reviews tracing such ideas from late 19th-century antecedents to recent work and further details on the interpretation of the skewness-versus-spread plot.

Let us close with an example for data on 158 glacial cirques from the English Lake District (Evans and Cox 1995). Glacial cirques are hollows excavated by glaciers that are open downstream, bounded upstream by the crest of a steep slope (wall), and arcuate in plan around a more gently sloping floor. More informally, they are sometimes described as "armchair-shaped". Glacial cirques are common in mountain areas that have or have had glaciers present. Three among many possible measurements of their size are length, width and wall height, and the distribution of all in the area studied is shown by

```
. skewplot length width wall_height, legend(ring(0) position(5) column(1))
```

to be markedly right skew (figure 8). Logarithmic transformation seems an obvious possibility, after which

---

3. Gilchrist calls the special case for quartiles Galton's skewness (pages 8, 25, 53, and 72), but there is no evidence that Galton used it.

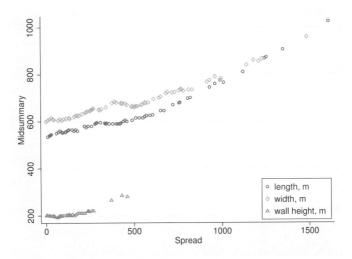

Figure 8: Skewness plot for three variables. The systematic upward drift indicates marked right skewness.

```
. skewplot log_length log_width log_wall_height, legend(ring(0) position(3)
>       column(1))
```

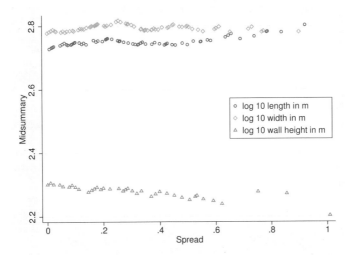

Figure 9: Skewness plot for three log-transformed variables. Approximately horizontal patterns indicate that transformations have yielded near symmetry of distributions.

shows approximate lack of skew (although a hint of mild overtransformation in wall height, which is best left alone for simplicity) (figure 9). An important feature of such plots is that the effects of outliers are localized.

7   Conclusions

With one command or another, users can now plot univariate distributions in many different ways. You can choose between several depictions of the density function or several depictions of the distribution function or its inverse, the quantile function. You can choose discrete or continuous representations and vertical or horizontal alignments. Less obviously, it is straightforward to add details (for example, rugs of distinct data values) or exploit the inbuilt flexibility of `graph` (for example, by looking at density estimates on a log scale or by constructing your own histogram with varying bin width).

The theme of distributions will continue into the next column but with a focus on categorical data. Distributions of categorical variables may be shown in a variety of displays: the survey will range from old staples to less well-known plots, with emphasis on the important special cases of graded data and of three variables with constant sum.

8   Acknowledgments

Ian Evans provided the Lake District cirques data, pointed me to the World Glacier Inventory data, and participated in many discussions on statistical graphics over more than 30 years. Marcello Pagano persistently urged the merits of equal-probability histograms and parenthetically underlined the connection with chi-squared tests. Vince Wiggins specifically alerted me to spanning bars and generally advised on strategies and tactics for using the new graphics. Elizabeth Allred, Ronán Conroy, Philip Ender, and Roger Harbord made helpful comments during development of various predecessors or versions of some programs discussed here. Richard Groeneveld kindly tracked down the Bowley reference.

9   References

Altman, D. G. 1991. *Practical Statistics for Medical Research*. London: Chapman & Hall.

Bagnold, R. A. 1937. The size-grading of sand by wind. *Proceedings of the Royal Society, Series A* 163: 250–264.

———. 1941. *The Physics of Blown Sand and Desert Dunes*. London: Methuen.

———. 1990. *Sand, Wind, and War: Memoirs of a Desert Explorer*. Tucson: University of Arizona Press.

Bardou, F., J.-P. Bouchaud, A. Aspect, and C. Cohen-Tannoudji. 2002. *Lévy Statistics and Laser Cooling: How Rare Events Bring Atoms to Rest*. Cambridge: Cambridge University Press.

Benjamini, Y., and A. M. Krieger. 1996. Concepts and measures for skewness with data-analytic implications. *Canadian Journal of Statistics* 24: 131–140.

————. 1999. Skewness—concepts and measures. In *Encyclopedia of Statistical Sciences Update*, ed. S. Kotz, C. B. Read, and D. L. Banks, vol. 3, 663–670. New York: Wiley.

Bowley, A. L. 1902. *Elements of Statistics*. 2nd ed. London: P. S. King.

Bowman, A. W., and A. Azzalini. 1997. *Applied Smoothing Techniques for Data Analysis: The Kernel Approach with S-Plus Applications*. Oxford: Oxford University Press.

Breiman, L. 1973. *Statistics: With a View Toward Applications*. Boston: Houghton Mifflin.

Chambers, J. M., W. S. Cleveland, B. Kleiner, and P. A. Tukey. 1983. *Graphical Methods for Data Analysis*. Belmont, CA: Wadsworth.

Cleveland, W. S. 1993. *Visualizing Data*. Summit, NJ: Hobart.

————. 1994. *The Elements of Graphing Data*. Rev. ed. Summit, NJ: Hobart.

Computing Resource Center. 1985. *STATA/Graphics User's Guide*. Los Angeles, CA: Computing Resource Center.

Cox, N. J. 1999a. gr41: Distribution function plots. *Stata Technical Bulletin* 51: 12–16. Reprinted in *Stata Technical Bulletin Reprints*, vol. 9, pp. 108–112. College Station, TX: Stata Press.

————. 1999b. gr42: Quantile plots, generalized. *Stata Technical Bulletin* 51: 16–18. Reprinted in *Stata Technical Bulletin Reprints*, vol. 9, pp. 113–116. College Station, TX: Stata Press.

————. 2001. gr42.1: Quantile plots, generalized: update to Stata 7.0. *Stata Technical Bulletin* 61: 10–11. Reprinted in *Stata Technical Bulletin Reprints*, vol. 10, pp. 55–56. College Station, TX: Stata Press.

————. 2003a. Software update: gr41_1: Distribution function plots. *Stata Journal* 3: 211.

————. 2003b. Software update: gr41_2: Distribution function plots. *Stata Journal* 3: 449.

————. 2003c. Stata tip 2: Building with floors and ceilings. *Stata Journal* 3: 446–447.

————. 2004. Software update: gr42_2: Quantile plots, generalized. *Stata Journal* 4: 97.

David, F. N., and N. L. Johnson. 1956. Some tests of significance with ordered variables. *Journal of the Royal Statistical Society, Series B* 18: 1–20.

Davison, A. C. 2003. *Statistical Models*. Cambridge: Cambridge University Press.

Evans, I. S., and N. J. Cox. 1995. The form of glacial cirques in the English Lake District, Cumbria. *Zeitschrift für Geomorphologie* 39: 175–202.

Fisher, N. I. 1983. Graphical methods in nonparametric statistics: A review and annotated bibliography. *International Statistical Review* 51: 25–38.

Freedman, D., R. Pisani, and R. Purves. 1978. *Statistics*. New York: W. W. Norton.

Galton, F. 1875. Statistics by intercomparison, with remarks on the law of frequency of error. *Philosophical Magazine*, 4th ser., 49: 33–46.

Gilchrist, W. G. 2000. *Statistical Modelling with Quantile Functions*. Boca Raton, FL: Chapman & Hall/CRC.

Gnanadesikan, R. 1977. *Methods for Statistical Data Analysis of Multivariate Observations*. New York: Wiley.

———. 1997. *Methods for Statistical Data Analysis of Multivariate Observations*. 2nd ed. New York: Wiley.

Groeneveld, R. 1998. Skewness, Bowley's measure of. In *Encyclopedia of Statistical Scienes Update*, ed. S. Kotz, C. B. Read, and D. L. Banks, vol. 2, 619–621. New York: Wiley.

Gumbel, E. J. 1943. On the reliability of the classical chi-square test. *Annals of Mathematical Statistics* 14: 253–263.

Hald, A. 1990. *A History of Probability and Statistics and their Applications before 1750*. New York: Wiley.

Hazelton, M. L. 2003. A graphical tool for assessing normality. *American Statistician* 57: 285–288.

Hoaglin, D. C. 1985. Using quantiles to study shape. In *Exploring Data Tables, Trends, and Shapes*, ed. D. C. Hoaglin, F. Mosteller, and J. W. Tukey, 417–460. New York: Wiley.

van Langren, M. F. 1644. *La Verdadera Longitud po Mar y Tierra*. Antwerp.

Mann, H. B., and A. Wald. 1942. On the choice of the number of class intervals in the application of the chi-square test. *Annals of Mathematical Statistics* 13: 306–317.

Parzen, E. 1979. Nonparametric statistical data modeling. *Journal of the American Statistical Association* 74: 105–131.

Quetelet, A. 1827. Recherches sur la population, les naissances, les décès, les prisons, les dépôts de mendicité, etc., dans le Royaume des Pays-Bas. *Nouveaux Mémoires de l'Académie Royale des Sciences et Belles-lettres de Bruxelles* 4: 117–192.

Salgado-Ugarte, I. H., and M. A. Pérez-Hernández. 2003. Exploring the use of variable bandwidth kernel density estimators. *Stata Journal* 3: 133–147.

Scott, D. W. 1992. *Multivariate Density Estimation: Theory, Practice, and Visualization*. New York: Wiley.

Silverman, B. W. 1986. *Density Estimation for Statistics and Data Analysis*. Boca Raton, FL: Chapman & Hall/CRC.

Simonoff, J. S. 1996. *Smoothing Methods in Statistics*. New York: Springer.

Stigler, S. M. 1986. *The History of Statistics: The Measurement of Uncertainty before 1900*. Cambridge, MA: Belknap Press.

Thorne, C. R., R. C. MacArthur, and J. B. Bradley, eds. 1988. *The Physics of Sediment Transport by Wind and Water: A Collection of Hallmark Papers by R. A. Bagnold*. New York: American Society of Civil Engineers.

Tufte, E. R. 1997. *Visual Explanations: Images and Quantities, Evidence and Narrative*. Cheshire, CT: Graphics Press.

Tukey, J. W. 1977. *Exploratory Data Analysis*. Reading, MA: Addison–Wesley.

Van Kerm, P. 2003. Adaptive kernel density estimation. *Stata Journal* 3: 148–156.

Wand, M. P., and M. C. Jones. 1995. *Kernel Smoothing*. London: Chapman & Hall/CRC.

Wild, C. J., and G. A. F. Seber. 2000. *Chance Encounters: A First Course in Data Analysis and Inference*. New York: Wiley.

Wilk, M. B., and R. Gnanadesikan. 1968. Probability plotting methods for the analysis of data. *Biometrika* 55: 1–17.

Wilkinson, L. 1999. Dot plots. *American Statistician* 53: 276–281.

About the author

Nicholas Cox is a statistically minded geographer at the University of Durham. He contributes talks, postings, FAQs, and programs to the Stata user community. He has also co-authored fourteen commands in official Stata. He was an author of several inserts in the *Stata Technical Bulletin* and is Executive Editor of the *Stata Journal*.

The Stata Journal (2004)
4, Number 2, pp. 190–215

Speaking Stata: Graphing categorical and compositional data

Nicholas J. Cox
University of Durham, UK
n.j.cox@durham.ac.uk

Abstract. A variety of graphs have been devised for categorical and compositional data, ranging from widely familiar to more unusual displays. Both official Stata commands and user-written programs are available. After a stacking trick for binary responses is explained, bar charts and related displays for cross-tabulations are discussed in detail. Tips and tricks are introduced for plotting cumulative distributions of graded (ordinal) data. Triangular plots are explained for three-way compositions, such as three proportions or percentages.

Keywords: gr0004, graphics, categorical data, binary data, nominal data, ordinal data, grades, compositional data, cross-tabulations, bar charts, cumulative distributions, logit scale, catplot, tabplot, tableplot, distplot, mylabels, triplot

1   Introduction

Given the new graphics introduced in Stata 8, *Speaking Stata* has turned to discuss graphics directly. In the previous column (Cox 2004), we started with the fundamental issue of graphing univariate distributions. We now focus on graphing categorical and compositional data, with particular emphasis on ways of going beyond what is obviously and readily available in official Stata.

William Cleveland, the author of two key graphics texts (Cleveland 1993, 1994), has suggested that "for 80% of all datasets, 95% of the information can be seen in a good graph" (Bentley 1988, 60). As far as many categorical datasets are concerned, one might wonder if they fell in the other 20% or if that good graph were proving extraordinarily elusive. Indeed, graphics and categorical data are not obvious bedfellows. A common caricature runs that, with measured data, you should start with simple graphs, while with categorical data, you should start with simple tables. But like even good caricatures, this picture is both true and false and needs qualification.

First, simple bar and pie charts of categorical data will have been familiar to many readers since childhood, but they are not covered by many introductory texts. Perhaps they are considered too elementary or too trivial. One good exception is Wild and Seber (2000). At another extreme, many categorical data analysis texts in the 1970s and 1980s stressed the use of log-linear and logistic models but paid little or no attention to graphical representation. On the other hand, several recent texts (e.g., Lloyd 1999; Agresti 2002; Simonoff 2003) are more concerned with emphasizing the strong links between categorical data analysis and other branches of statistics. For that and other reasons, such texts pay greater attention to graphics. In addition, and going beyond what can be considered here, there is a substantial but fairly self-contained literature on biplots and related displays, many of which are designed for categorical data (Gower and Hand 1996; Blasius and Greenacre 1998; Friendly 2000). (Ulrich Kohler has a Stata program called `biplot` on SSC.)

In this column, we will start with a simple stacking trick for binary responses. Then we will look in more detail at bar charts and variations on them. Simple they may be, but in many ways they remain among the most effective graphics for categorical data. Then we will examine some displays that perform well for ordinal data, specifically graded data. Finally, we will look at a standard triangular plot for compositional data, specifically for the case of three categories whose percentages necessarily sum to 100, allowing a two-dimensional representation.

There are two key references throughout this column. The first is the *Stata Graphics Reference Manual* [G]; the second is the most useful and complementary compendium of examples given by Mitchell (2004).

2 Kinds of categorical data

Let us first review some terminology: depending on your field, several terms are likely to appear standard, but a few are nonstandard. Many texts classify kinds of data, but those classifications seem to coincide only when one text cites another. However, several classifications mix together cross-combinations of (1) discrete versus continuous (a standard mathematical distinction); (2) nominal, ordinal, interval, and ratio (a scheme introduced by Stevens 1946); and (3) fuzzier distinctions, such as categorical versus measured and qualitative versus quantitative (authors differ on where to draw the line).

Categorical data are regarded here as those for which the original raw data are qualitatively distinct categories, rather than quantitative measurements. Those raw data may be quickly converted into quantitative codes, counts (frequencies) over categories, or proportions or percentages of occurrence of categories. They may commonly be received by users in some such converted form.

Binary (Boolean, dichotomous, quantal) data take only two states, which are often coded 0 and 1. They will be very familiar to almost all readers, common examples being yes or no, alive or dead, success or failure, and so forth. Polytomous or multistate data (often loosely called nominal) may take on three or more states, but those states lack a natural order. In practice, many of the often-quoted examples are not clear-cut, partly because of the temptations of giving any scale an ordered interpretation. Thus, marital status (single, married, divorced), occupations, colors, diseases, and political parties can often be ordered, at least approximately, from some viewpoint.

Ordinal data may take on several states, but those states do have a natural (meaning uncontroversial) order. In practice, ordinal data span an enormous range. At one extreme, an ordinal variable may take on a relatively small number of ordered categories (sometimes called 'grades'), such as opinions on some defined scale from 'Strongly agree' to 'Strongly disagree'. Then there are ranks, which are, in the strict sense, permutations of integers 1 up and so possess much structure (Diaconis 1988; Marden 1995). From another viewpoint, ranks are akin to counts. We must also consider test scores or other composite scales (e.g., assessments of disability or pain) that are often treated as if they were continuous measurements to some approximation. (Where would educational systems be without grade-point averages or the equivalent?)

Graphically, these differing kinds of categorical data present unequal challenges. Binary data often invite—indeed demand—a reduction to probabilities of one or other possible outcome, which can then be plotted directly. The main difficulty may then lie in showing data points that are all either 0 or 1, or more generally, just one of two possible values. However, one simple and perhaps unfamiliar technique is presented here to begin with. At the other extreme, the closer ordinal data are to quantitative scales that are nearly continuous, the less they need any special form of graphical treatment. The main needs for distinctive graphical forms therefore arise with polytomous or coarse ordinal scales.

3   A stacking trick for binary responses

Suppose that we have a binary response (coded 0 and 1 in most examples) and that an important covariate also shows some discreteness, if only as a consequence of the resolution of recording. In the `auto` dataset,

```
. sysuse auto
```

one such problem is illustrated by figure 1:

```
. scatter foreign mpg, ylabel(0 1, valuelabel angle(h))
```

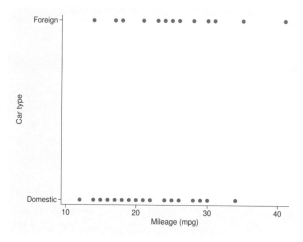

Figure 1: The relationship between car type and mileage is obscured by overplotting of identical data points.

This graph shows many fewer plotted points (30) than there are observations (74) because of ties on both variables, but we have no way of reading any frequencies off the graph. Such ties will often occur with many kinds of data, as with, say, patients' ages in years or incomes recorded in rounded terms. You may already know a solution to the graphical problem: use the `jitter()` option. Jittering does not always work very well, especially if you want a graph for public use. It may also require more explanation in print (or more argument with reviewers) than the trick to be explained now, which is a `dotplot`-like idea of little bars of points stacked vertically. Here is one recipe. As with all demonstration cookery, the details have been customized slightly so that the example works well, but the main idea is easy to grasp.

What is the largest number of ties? We can read that from the results of `tabulate mpg foreign` or calculate it directly by

```
. bysort foreign mpg : generate freq = _N
. summarize freq
```

which in our example shows a maximum of 8. (This last technique may require more care in the presence of missing values, not an issue here.) Let us decide that we want the longest bars to be of length 0.1. We can now produce a variable to be plotted in a single command:

```
. bysort foreign mpg : generate foreign2 =
> cond(foreign == 1, 1 - 0.1 * (_n - 1)/7, foreign + 0.1 * (_n - 1)/7)
. label values foreign2 origin
```

However, that single command does deserve some explanation. The `cond()` function specifies what is to be done both when `foreign == 1` is true and when that is false. The `foreign +` is clearly superfluous when `foreign` is 0, which is the only other value occurring in the `auto` data. For a more general solution, we should worry about dealing properly with missing values, which will map to missing whenever anything is added.

Within each group of tied values on `foreign` and `mpg`, the built-in variable `_n` varies from 1 to 8 (as just ascertained), so $(\_n - 1)/(\_N - 1)$ in general and $(\_n - 1)/7$ in our example vary from 0 to 1. Thus, in constructing a variable to be plotted, we subtract (at most) 0.1 from 1 and add (at most) 0.1 to 0. The details to tune according to taste and circumstance are thus (a) the constant, which is 0.1; (b) the maximum frequency −1; and (c) whether bars are added to or subtracted from the horizontal reference lines for 0 and 1.

```
. scatter foreign2 mpg, ylabel(0 1, valuelabel ang(h))
> ytitle(`"": variable label foreign`"")
```

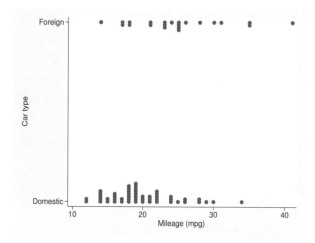

Figure 2: Data points are piled in little stacks to show frequencies of combinations.

Figure 2 shows the result. Note a small flourish here, automating the use of the variable label. As it happens, with this dataset just typing it in would take no more effort, but a general method would be preferable in a do-file or program that you might write for wider application.

4   Bar charts of frequencies

One of the simplest kinds of plots, but one often requested, is a bar chart of frequencies of one or more categorical variables. If we consider what is available in official Stata, the natural commands to consider here are `histogram` and `graph bar` or `graph hbar`.

4.1   Using histogram

`histogram` is optimized for the case of continuous variables; by default, you get a series of touching bins. You can spell out that you want a graph emphasizing a discrete scale by adding the `discrete` option. You can specify gaps between bars, which are customary with categorical scales to show that the scale really is *not* continuous, by using an option such as `gap(50)`. You can insist that value labels, likely to be attached to most categorical variables, be shown by using another option, `xlabel(, valuelabel)`. These tweaks may be sufficient for some variables, but in total, `histogram` may remain problematic for categorical scales.

First, several lengthy value labels may be difficult to read when placed side by side along a horizontal axis. Changing the font size or the orientation of the text may just replace one problem with another: at one extreme, we approach giraffe graphics in which readers are presumed able and willing to move their heads over a range of angles to scrutinize the plot presented. However, using the `horizontal` option could be sufficient to solve this.

Second, many designers of bar charts and perhaps even more readers prefer a display more colorful than the monochrome display standard with histograms.

Finally, `histogram` can be extended to a two-variable classification, as a multiple panel display is obtainable with the `by()` option. Sometimes this works well, but equally sometimes you would prefer something different.

4.2   Using graph bar or graph hbar

`graph bar`, `graph hbar`, and, for that matter, `graph dot` offer a different approach. It is easiest if the frequencies come predefined as a variable because each command can then be used with either (`asis`) or (`sum`). But if you want Stata to do the counting for you, you must do things another way. In particular, with the `auto` data read in,

    . graph hbar (count) rep78

does not give you the frequencies of the categories of `rep78`. It counts nonmissing values of `rep78` and shows a single bar of height 69. Moreover,

    . graph hbar (count) rep78, by(rep78)

is illegal. More positively, something like

    . graph hbar (count) mpg, by(rep78)

does what you want but at the cost of a mislabeled graph and some strain on the brain. Incidentally, being able to say why this works as a bar chart of the frequencies of `rep78`, apart from the mislabeling, is a little test of your understanding of what is going on.

The underlying problem here is that `graph hbar` and its siblings `graph bar` and `graph dot` are built around a temporary `collapse` of the data, whereas in effect what we want here is a temporary `contract`. However, that problem is a little difficult to diagnose without looking inside the graphics code.

Another way to do it is by calculating something in advance, as in

    . generate freq = 1
    . graph hbar (count) freq, over(rep78)

and yet another way to do it is to `contract`, as in

    . contract rep78
    . graph hbar (asis) _freq, over(rep78)

The first of these may smack of a programmer's trick, while the second has the strong disadvantage that the original dataset will have to be read in repeatedly to do a series of bar charts or even to return to the original data for further analyses.

Arguably, none of these solutions is good, especially when we consider the common need for also showing fractions or percentages. In addition, we would expect a bar chart command to handle smoothly any missing values or `if` or `in` conditions.

4.3   Using catplot

`catplot` is a convenience command designed to avoid these awkwardnesses. The aim is simply that a bar or dot chart of frequencies be available through a single command, without any need for preparation or restructure of the dataset. `cat` is meant to suggest category, naturally, but any feline undertones should be regarded as felicitous. You may install `catplot` from SSC using the `ssc` command ([R] **ssc**):

```
. ssc install catplot
```

To be precise, `catplot` shows frequencies (or, optionally, fractions or percentages) of the categories of one, two, or three categorical variables. The first named variable is innermost on the display so that its categories vary fastest along the axis. The syntaxes `catplot bar`, `catplot hbar`, and `catplot dot` indicate use of `graph bar`, `graph hbar`, and `graph dot`, respectively. The choice is a matter of personal taste, although in general horizontal displays make it easier to identify names or labels of categories and so avoid the giraffe graphics just deprecated.

The default display with `bar` and `hbar` is graphically conservative, reflecting the view that height of bars and text indicating categories are the best ways of conveying information. If you wish also to have bars in different colors, specify the option `asyvars`, which differentiates the categories of the *first* named variable; to stack bars of different colors, specify the further option `stack`.

The default display with `dot` is similarly conservative. If you wish to have point symbols in different colors, similarly specify the option `asyvars`, which differentiates the categories of the *first* named variable; to use different point symbols, use the further option `marker()`.

There are various simple options to show percentages or fractions. `percent` indicates that all frequencies should be shown as percentages (with sum 100) of the total frequency of all values being represented in the graph, while `percent(`*varlist*`)` indicates that all frequencies should be shown as percentages (with sum 100) of the total frequency for each distinct category defined by the combinations of *varlist*. There are similar fraction options for fractions with sum 1.

The `sort` option specifies that values shown should be sorted in each category (higher values at the bottom of each category). Sorting is applied to all variables shown. The `descending` option specifies that sorted values should be shown in descending order (higher values at the top of each category).

Simple examples with just one categorical variable are

```
. catplot hbar rep78, sort
. catplot hbar rep78, sort desc
```

and with two such variables are

```
. catplot hbar rep78 foreign
. catplot bar rep78, ylabel(, angle(h)) percent(foreign)
> by(foreign, note("") subtitle(Repair record 1978, pos(6)))
```

```
. catplot hbar rep78 foreign, percent(foreign) blabel(bar, position(outside)
> format(%3.1f)) ylabel(none) yscale(r(0,60)) ytitle("")
> subtitle(Repair record 1978 distribution by car origin)
```

The last two of these are shown as figure 3. Almost all the flexibility comes from the `graph` commands for which `catplot` is a wrapper.

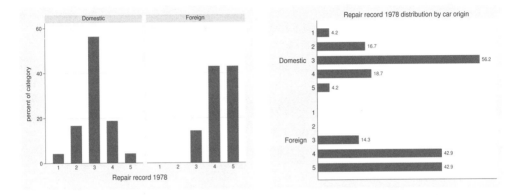

Figure 3: `catplot` can produce bar charts of percentages.

4.4 Hybridizing graphs and tables

A step away from the displays afforded by `catplot` are what may be called graphical tables. Typically, the value in each cell of such displays is represented by a bar, a spike, an interval, or some other similar way. The commands `tabplot` and `tableplot` from SSC provide some flexibility here. Both are applicable to problems with two categorical variables and some third quantity to be plotted for each cross-combination of categories.

`tabplot` is specialized to show counts (or optionally percentages or fractions of total, rows, or columns) and to show vertical or horizontal bars. For the most part, it can be considered as a wrapper program for `twoway rbar`. The name is intended to remind you of `tabulate`, which many users happily abbreviate `tab`, a command centered on showing tables of counts (or percentages). `tabplot` works best in practice when at least one variable is ordinal so that intelligible patterns might be expected in relation to at least one axis. By tuning the `barwidth()` option, the bars can be made very thin, which is not especially useful, or they can be made so wide that they touch, thus giving a set of juxtaposed histograms, which can be very useful. `tabplot` is in fact not restricted to categorical variables; the variable on either axis or the variables on both can be flagged as to be treated literally. Hence, `tabplot` has a secondary use for juxtaposing histograms, but that is not explored fully here.

`tabplot` is primarily intended to show the main structure of a table. It can be useful for exploratory inspection or even public presentation of moderate-sized tables, say, those with about 10 or 20 rows or columns. At that size, even experienced analysts can sometimes find it difficult to see the table for the cells. On the other hand, `tabplot` gives up on showing detailed numeric scales on either axis. There is just one token concession to showing numbers as such—a note indicates the largest value shown in any cell. Beyond that, `tabplot` relies on users' mental assessment of bar or spike heights or lengths. Fortunately, this is the easiest graphical skill (Cleveland 1994) so that fine as well as coarse structure can usually be discerned. For table look-up of individual cell magnitudes, there is, unsurprisingly, nothing superior to table look-up. More simply put, the graphs produced by `tabplot` and the equivalent tables complement each other. In practice, many thesis advisors and journal editors are reluctant to approve publishing both on the grounds that they show the same information.

An example is provided by a classic dataset (Tocher 1908) on children's eye and hair color in Caithness in northern Scotland, which has been analyzed in various ways in the statistical literature (e.g., Fisher 1940). Historically Caithness has experienced influxes of migrants not only from other parts of Britain but also from other areas, Scandinavia in particular, so its people show very interesting genetic diversity. Both variables may be regarded as ordinal, at least approximately.

A table comes as usual from

```
. tab eye hair [fw=freq]
```

and a basic frequency plot (figure 4) is then given by

```
. tabplot eye hair [fw=freq], horizontal ytitle(, orient(hor))
```

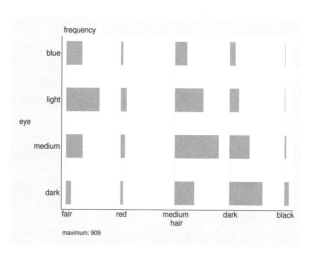

Figure 4: `tabplot` can produce graphical displays of two-way tables of frequencies.

The syntax evidently is designed to resemble that of `tabulate`. In each command, the first-named variable goes on the left of the display (the rows of the table, the vertical axis of the graph) and the second-named variable on the bottom of the display. Note also that, by default, the lowest-numbered row (whenever the row variable is numeric, which is usual but not essential) is shown at the top of the vertical axis.

When all values shown (here the frequencies) are zero or positive, the maximum bar height or length is, by default, 0.8. As the interval between successive categories is, by default, 1, these choices normally imply that neighboring rows and columns remain distinct.

One key scaling allowed is to percentages of either row or column categories. In addition, scaling to fractions is possible, but merely cosmetic in affecting only marginal text and the note about maximum value. One example is thus (figure 5):

```
. tabplot eye hair [fw=freq], horizontal percent(eye) ytitle(, orient(hor))
```

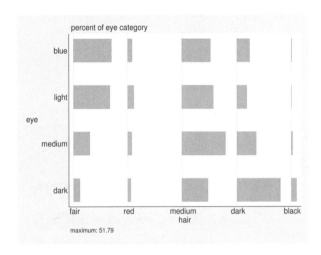

Figure 5: `tabplot` also can produce graphical displays of two-way tables of percentages, here for row categories.

The broad association between eye and hair color should come as no surprise even to rank amateurs in genetics or physical anthropology. There is a diagonal pattern stretching from fair-haired and blue-eyed to black-haired and dark-eyed. Which variable is taken as base for percentages appears arbitrary; experts might be able to advise. In your own field, it will usually be clearer which kind of display is best for your problem and dataset. In this example, we try vertical bars and column percentages (figure 6):

```
. tabplot eye hair [fw=freq], percent(hair) ytitle(, orient(hor))
```

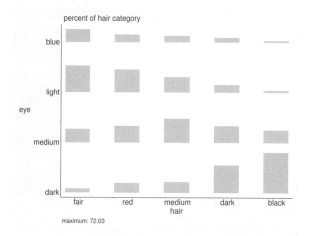

Figure 6: As a variation, `tabplot` here produces vertical bar displays of percentages for column categories.

Let us compare this kind of display with a stacked, or divided, bar chart. Stacking is available through `graph bar` or `graph hbar` or through commands such as `catplot` (figure 7):

```
. catplot hbar eye hair [fw=freq], percent(hair) stack asyvars
> bar(1, bcolor(gs14)) bar(2, bcolor(gs10)) bar(3, bcolor(gs6)) bar(4, bcolor(gs2))
```

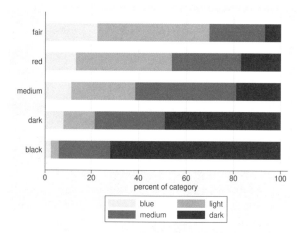

Figure 7: Stacked (divided) bar charts are widely familiar, but only the end bars are anchored to straight base lines, inhibiting the decoding of intermediate category values.

Here, as a small but telling detail, we spell out that bar colors are to be various gray scales (e.g., `gs14`). The main role of color, arguably, is to provide qualitative contrasts.

Nevertheless, it is worth checking that, as far as possible, colors representing ordinal scales do represent a monotonic sequence.

Stacking of percentages, so that divided bars of constant length each represent group totals, can serve as a useful reminder to inexperienced readers of the basis of the graph. Alternatively, it might seem an uninformative tautology: "This graph's main message is that all percentages sum to 100. So, what else is new? I am obliged to look inside each bar for the real information in the data." Naturally much hinges on what you are trying to do and who you have in mind as readers of your graph.

A pattern common with frequencies reduced to percentages is that the percentages in extreme categories behave monotonically, but this is often not true of intermediate categories. Thus from one end of each axis to the other, the percentage of dark or black goes up while the percentage of fair or blue goes down, while other categories sometimes change monotonically and sometimes change through a turning point. All this is no surprise on reflection, but a distinct weakness of the stacked design is that only the bars for extreme categories are based on clear reference lines. It is especially difficult to decode values of intermediate categories accurately; indeed, perhaps few readers try very hard to do this. Clearly, one key advantage of displays like those from `tabplot` is that bar heights, and thus values in the table, may be compared more easily, at least within rows or within columns, because all bars are based on reference lines.

There are other ways of tackling this issue. One is sliding each divided bar so that it straddles a central category. The program `slideplot` on SSC is dedicated to such sliding bar plots, which sometimes work well. Unfortunately, they can also create as many problems as they solve. Nevertheless Stouffer et al. (1949) provided one source with plentiful examples in their monograph, one of the classics of quantitative social science.

`tableplot` is more general than `tabplot`, but to get what you want, you may need to do more work. The main idea, as with `tabplot`, is that two variables provide a framework of rows and columns defining cells. Unlike with `tabplot`, a third variable must be specified. Its values must be unique within each cell defined by the rows and columns. In addition, a range of plot types is available, namely `rbar`, `rcap`, `rcapsym`, and `rspike`. Thus, you can see that `tableplot` is mostly a wrapper for `twoway` with one of the named plot types. In practice, most users seem to refer `rbar`.

The name `tableplot` is intended to be reminiscent of `table`, a fairly general tabulation command, but one for which it is more common to specify precisely what is tabulated. The analogy is not strict and should be promptly forgotten if it does not seem helpful.

Thus the main syntax is schematically `tableplot` *plottype showvar rowvar colvar*. You may also think of this as *plottype z y x*: here y and x define rows and columns, as with `tabplot`, but the third variable, z, gives the heights of the bars, spikes, or intervals plotted within each cell. For example, therefore,

```
. tableplot rbar freq eye hair
```

is equivalent to

```
. tabplot eye hair [fw=freq]
```

Using `tableplot` to do what `tabplot` can do directly would be missing the point, however. More interesting applications go beyond representing tables of counts or percentages. An example dataset comes from a study by Townsend (1995) of various pioneer communities in the Mexican rainforest. In several villages, her team asked how many men and how many women did various tasks, such as doing laundry, cooking, fetching water, fetching wood, and so forth. Their data can be summarized, for each task and each village, on a scale showing distribution of labor by gender:

(number of women − number of men) / (number of women + number of men)

So, if no men do some task, this measure is 1; if no women do it, it is −1; and if equal numbers of men and women do it, it is 0. Depending on your background, you may recognize this as a difference in probabilities, and thus, as similar to several statistical measures or as equivalent to a fractional majority for elections. No doubt other applications exist. Incidentally, note that, in this dataset, the ratio of numbers of men and of women and its reciprocal are both impractical: zeros for some places, some tasks, and both genders would render both ratios indeterminate. Even if that were not true, a serious limit to the usefulness of any sex ratio is its skewness around 1. The scale just defined with its symmetric limits is greatly preferable. The `tableplot` just below required some prior work sorting row and column categories into the best order. Even though neither task nor place is an ordinal variable, sorting rows according to, say, medians across columns and vice versa proved very helpful. More generally, this is often one of the most fruitful steps to identifying underlying structure.

Figure 8 shows the gradation from tasks generally done by women (laundry, cooking, care of hens, fetching water) to those generally done by men (preparing land, harvesting, care of cows, fetching wood). Interestingly, the two middle tasks differ: gardening and milking cows both average about an equal division of labor. In the case of gardening, this is fairly consistent between places, but with milking cows, there is much more variability between places. Subject-matter experts might be able to comment, but the graphical point is simple. This plot represents a 10×12 table and allows both coarse and fine structure to be seen. The decision to put tasks on the rows and places on the columns is clear-cut, given the nature of variability in the data. There is much more variation between tasks than between places, as could be explored further in a formal analysis, if it were thought fit to treat the data as if they were a random sample. Also, the tasks are easy even for nonspecialists to think about, but the place names mean little except to those familiar with the area. Being obliged to put the names vertically does supply an instance of giraffe graphics, but it is the easiest sacrifice to be made. Other solutions, such as abbreviating the names, writing them at 45°, or using a smaller font size, would here gain nothing.

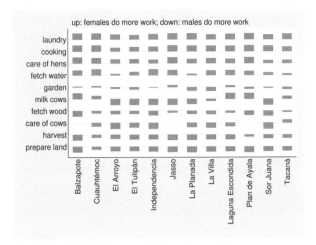

Figure 8: `tableplot` permits the display of values of a third variable by combinations of two variables. Here the division of labor in some Mexican villages between females and males is shown for various tasks.

5   Cumulative distributions for graded variables

A very different approach for ordinal data, especially graded variables, is based on another standard kind of graph, showing cumulative distribution functions. `distplot` (Cox 1999, 2003a,b, 2004) produces plots of cumulative distribution functions or their reverses.

One particular option was introduced with graded data specifically in mind and is especially appropriate for showing data with a relatively small number of categories. `midpoint` specifies the use of midpoints of probability intervals for each distinct value so that the cumulative probability P for a variable X is defined as

$$\Pr(X < x) + \frac{1}{2}\Pr(X = x)$$

With terminology from Tukey (1977, 496–497), this could be called a 'split fraction below'. It is also a 'ridit' as defined by Bross (1958); see also Fleiss, Levin, and Paik (2003, 198–205) or Flora (1988). Yet again, it is the mid-distribution function of Parzen (1993, 3298) and the grade function of Haberman (1996, 240–241).

Using this definition rather than $\Pr(X < x)$ or $\Pr(X \leq x)$ means that more use is made of the information in the data. Either alternative would always mean that some probabilities are identically 0 or 1, which tells us nothing about the data. In addition, there are fewer problems in showing the cumulative distribution on any scale for which the transform of 0 or 1 is not plottable. This approach for graded data was first implemented in Stata by Cox (2001). Its roots go back at least to Tukey (1961), a paper that was, however, not published until 1986.

To develop that point, `distplot` has an option useful for graded data. `trscale()` specifies the use of an alternative transformed scale for cumulatives. Stata syntax should be used with @ as a placeholder for untransformed values. So, to show probabilities as percentages, specify `trscale(100 * @)`; on an inverse normal scale, specify `trscale(invnorm(@))`; on a logit scale, specify `trscale(logit(@))`; and on a cloglog scale, specify `trscale(cloglog(@))`.

Further information on transformations for probability scales is available in Tukey (1960, 1961, 1977), Atkinson (1985), Cox and Snell (1989), and Emerson (1991). Some of the possible transformations appear as link functions in the literature on generalized linear models (e.g., McCullagh and Nelder 1989; Aitkin et al. 1989).

Let us look at some examples with this little toolkit in mind. For Stata users, repair record `rep78` in the `auto` data is a simple and familiar example of a graded variable. The first step is to look at plain cumulatives (figure 9):

```
. distplot connected rep78, ylabel(, angle(h)) midpoint by(foreign)
```

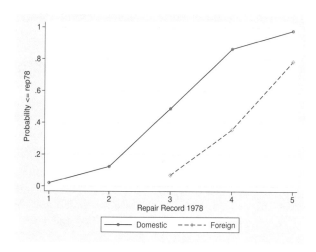

Figure 9: Cumulative distributions of graded variables can be shown using the midpoints of intervals of probability.

The fairly standard S-shape of the cumulative distribution for domestic cars particularly suggests that a transformed scale might yield a complementary view. I like using logits here, for no special reason except that they often work very well, as further examples will show. We can do that on the fly (figure 10):

```
. distplot connected rep78, ylabel(, angle(h) midpoint by(foreign)
> trscale(logit(@)) l2(logit scale)
```

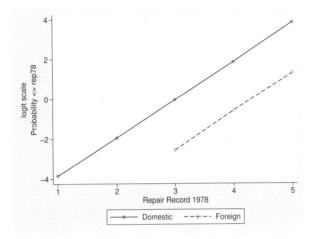

Figure 10: Logits of cumulative distributions of graded variables often plot as nearly straight lines, giving a handle on the difference a covariate makes.

The linearization is dramatic. The first time I got this, I suspected that a silly bug of mine was giving me lines as artifacts. Yet it is genuine. The more general heuristic is simple. First is the empirical observation that logits of cumulatives are often fairly straight. Note that this method is not, in fact, a back-door way of fitting one of the standard logit-based models to ordinal responses. The straightness seems to arise partly from how nature (or society) works and partly from how people devise grading schemes. For example, most schemes are devised so that some occurrences are expected in all categories. Second, do covariates make much of a difference, as shown by very different cumulatives for each group, and if so, is the ordering and even magnitude of effects what would be expected? Third, can extra structure be identified, such as highly anomalous groups or simple interaction effects?

Beyond that heuristic, we need to focus on one tacit assumption that for some will be troubling. Deciding whether cumulatives are straight or some other shape depends on the metric on which graded variables are being shown. The curve shape is contingent, in this example, on taking the values of rep78 quite literally (meaning, numerically). According to the purists, as represented by Stevens (1946) among others, this is precisely what you should not do with graded data. According to the pragmatists, on the other hand, you should feel free to do whatever works.

In this issue, I tend to side with the pragmatists, despite recognizing the purists' argument. It is always open to users to experiment with different scoring schemes outside of distplot. Moreover, if a graph does not help, you just should shrug your shoulders and move on to try other kinds of display. distplot with options midpoint and trscale() has been used to show simple patterns that do not need the gloss of a more formal analysis (Bentley et al., forthcoming). If it helps to select covariates for a modeling exercise, that also can be benefit enough.

Enough arm waving; let us look at some more examples. Fienberg (1980, 54–55) reported data from Duncan, Schuman, and Duncan (1973) from 1959 and 1971 surveys of a large U.S. city asking, "Are the radio and TV networks doing a good job, just a fair job, or a poor job?" Suppose that, underneath the labels below, `opinion` runs 1/3. `group` here evidently is a cross-combination of year and race, created off stage by the `egen` function `group()`. Mapping two or more covariates to one is a standard device here, so long as the total number of cross-combinations remains manageable.

```
    group       opinion      freq
 1. 1959 Black  Good          81
 2. 1959 Black  Fair          23
 3. 1959 Black  Poor           4
 4. 1959 White  Good         325
 5. 1959 White  Fair         253
 6. 1959 White  Poor          54
 7. 1971 Black  Good         224
 8. 1971 Black  Fair         144
 9. 1971 Black  Poor          24
10. 1971 White  Good         600
11. 1971 White  Fair         636
12. 1971 White  Poor         158
```

With these data, we will go straight to logits (figure 11):

```
. distplot connected opinion [w=freq], ylabel(, angle(h)) midpoint by(group)
> trscale(logit(@)) xlabel(1/3, valuelabel) l2(logit scale) legend(col(1)
> position(5) ring(0))
```

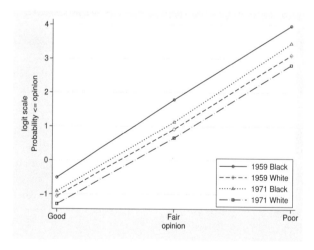

Figure 11: Opinion of radio and TV networks varies with race and has shifted over time, with a narrowing gap between black and white. Higher curves show collectively more favorable opinions.

This shows a clear shift of opinion towards Poor from 1959 to 1971 and a narrowing gap between Black and White. Otherwise said, race and year both make a difference, which seems no surprise. Whether the narrowing gap is modeled with some kind of interaction effect is an issue for more formal analyses.

As a third and final example, Knoke and Burke (1980, 68) gave data from the 1972 U.S. General Social Survey on church attendance. Suppose that, underneath the labels below, `attend` runs 1/3.

```
     group      attendance      freq
 1.  young non-Catholic     low      322
 2.  young non-Catholic  medium      122
 3.  young non-Catholic    high      141
 4.    old non-Catholic     low      250
 5.    old non-Catholic  medium      152
 6.    old non-Catholic    high      194
 7.       young Catholic     low       88
 8.       young Catholic  medium       45
 9.       young Catholic    high      106
10.         old Catholic     low       28
11.         old Catholic  medium       24
12.         old Catholic    high      119
```

The `reverse` option ensures that higher attendance groups plot farther to the right on the graph. There are clear age and denomination effects and an indication of an interaction between the two (figure 12):

```
. distplot connected attendance [w=freq] , ylabel(, angle(h)) by(group)
> midpoint trscale(logit(@)) legend(column(1) position(1) ring(0))
> xlabel(1/3, valuelabel) l2(logit scale) reverse
```

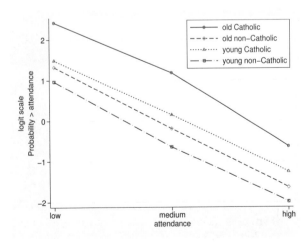

Figure 12: Church attendance is affected by age and denomination. Logits of cumulatives are again nearly straight. On this scale, there appears to be an interaction effect.

6   Better labels on transformed scales

The ability to produce axes on transformed scales may seem a mixed blessing. It raises an issue which is more general. We are happy thinking on a probability scale (about say 0.9, 0.42, 0.007) but might well be less happy thinking on a logit scale (about the equivalents, 2.197, −.323, −4.955 to 3 decimal places). There is a way of getting the best of both worlds. A transformed scale and also labeling in more intelligible terms are possible by using the `ylabel()` and `xlabel()` options to specify text to be shown at axis positions. (By the way, value labels are not general enough to work well, as they can only be attached to integers.)

Suppose that one graph axis is a logit scale, but you wish the axis labels to show untransformed probabilities. Stata could be used as a convenient calculator to work out the mapping, but a dedicated utility is preferable.

The idea behind `mylabels`, which may be downloaded from SSC, is that you feed it the numeric labels to be shown and the transformation being used. It will then place the appropriate specification in a local macro that you name. You may then use that local macro as part of a later graph command. A similar idea may be used for axis ticks: the command is called `myticks` and comes bundled with `mylabels`. The idea behind these programs may be traced to Royston (1996).

The option `myscale()` specifies the transformation. Stata syntax should be used with @ as placeholder for the original value. Hence, to show original values on a logit scale, specify `myscale(logit(@))`.

The option `local(macname)` inserts the option specification in local macro *macname* within the calling program's space. If you are unfamiliar with the idea of local macros in Stata, see [U] **18.3 Macros** or Cox (2002). The key idea is, essentially, to put all the definitions together in a bag which can then be referred to concisely. The macro will be accessible after `mylabels` or `myticks` has finished for subsequent use with `graph` or other graphics commands.

For example,

```
. mylabels 0.1(0.1)0.9, myscale(logit(@)) local(myla)
```

means that you have data on a logit scale but wish labels to be displayed that show values from 0.1 in steps of 0.1 to 0.9. What `mylabels` will show is

```
-2.19722 ".1" -1.38629 ".2" -.847298 ".3" -.405465 ".4" 0 ".5" .405465 ".6"
.847298 ".7" 1.38629 ".8" 2.19722 ".9"
```

That is, the text ".1" will be shown at −2.19722 on whatever axis is specified, the text ".2" at −1.38629, and so forth. The main point of showing you the list is allowing you to check that you have what you want.

On a graph, you may want plain ticks in between labels. `myticks` creates the list for you.

```
. myticks 0.15(0.1)0.85, myscale(logit(@)) local(myti)
```

Then you can re-issue your graph call (figure 13):

```
. distplot connected attendance [w=freq], ylabel(`myla´, ang(h)) by(group)
> midpoint trscale(logit(@)) legend(col(1) position(1) ring(0))
> xlabel(1/3, valuelabel) l2(logit scale) reverse
```

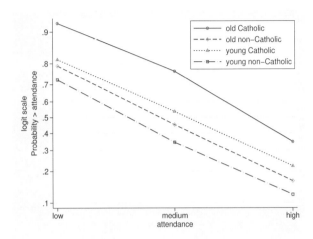

Figure 13: `mylabels` has been used to produce more intelligible labeling of the logit scale in terms of probabilities.

7 Triangular plots for three-way compositional data

Suppose that we have two variables that have a constant sum, such as $p =$ proportion female and $q =$ proportion male, so that $p + q = 1$. A plot of data for those variables shows all points lying on the line defined by that constraint. In practice, we would not draw that plot, unless by accident. We would recognize a bivariate situation as essentially univariate and examine the distribution of, for example, p.

More interesting is the case of three (zero or positive) variables with a constant sum, say three proportions with sum $p + q + r = 1$. This constraint defines a plane in (p, q, r) space. In fact, data points are confined to a triangular subset of that plane, which can thus be laid flat in two dimensions with no loss of information. The trivariate situation is essentially bivariate. Naturally this is just a special case of the more general situation of compositional data (Aitchison 1986), but it is nevertheless an interesting and frequent special case.

`triplot` from SSC produces a triangular plot of three variables, which are plotted on the left, right, and bottom sides of an equilateral triangle. Each should have values between 0 and some maximum value (default 1), and the sum of the three variables should be equal to that maximum (within rounding error). Most commonly, three fractions or proportions add to 1, or three percentages add to 100.

Triangular plots appear under various names in the literature, including percentage or reference triangles and barycentric, mixture, ternary, trilinear, or triaxial plots. Beyond this profusion of terminology there lies a curious pattern of patchy invention and use. Triangular plots have been rediscovered many times over in several fields yet are also apparently little or never used in some other fields, despite the potential for many applications. One root is the barycentric calculus of Möbius (1827); see Gray (1993). Other roots are 18th- and 19th-century studies of color mixing, photoelasticity, and the phase rule (Howarth 1996). Disciplines in which they are popular include genetics (for three genotypes or three alleles), geology (various geochemical compositions and particle shape analysis), pedology (for clay, silt, and sand fractions of a soil; for an early Stata implementation, see Danuso 1991) and political science (election data, often for two big parties and a bundle of others).

A common kind of economic example of triangular plots is based on some three-fold division of activities, say into agriculture (and sometimes other so-called primary activities, such as fishing, forestry, and mining); manufacturing (secondary); and services, including information (tertiary). This classification is some decades old, and some would now prefer to split services and information, producing four sectors, the last occasionally called quaternary. Such a scheme would take the classification beyond the reach of triangular plots. Nevertheless, the trichotomy remains of use for broad-brush description, especially for comparing a range of less developed and developed economies. Some World Bank data on the composition of Gross Domestic Product for 112 economies were used to produce figure 14:

```
. triplot agriculture industry services, separate(Africa) max(100) legend(pos(2)
> ring(0) col(1)) ms(O Th) note(units are %)
```

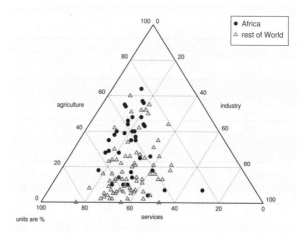

Figure 14: `triplot` is used here to show the structure of Gross Domestic Product in 112 economies in 1996 from World Bank data.

A detail here that may puzzle is the placement of the legend at `pos(2)` and within `ring(0)`. The explanation follows from the kind of Stata graph `triplot` is, underneath its skin. Despite appearances, the triangular plot is just a `twoway` plot with y and x axes removed. The internal grid, the triangular frame, and an optional Y within the triangle are defined by calls to `twoway connect` and `twoway line`, while all else is done by calls to `twoway scatter`. The clumsiest part is the programming of the axis labels, which do not have the flexibility of standard y or x axis labels. The nicest part is that almost all the handles you need are available as standard `twoway` features.

To explain the legend puzzle, the legend is within the plot region defined by the y and x axes. Those axes have been removed, but they nevertheless retain a shadowy existence. Other details of the `triplot` call should be more transparent. The `separate()` option subdivides data points into groups, here according to a binary variable (African or not).

A second example looks at how such compositions change over time. Given data on the composition of the civilian labor force in the United States (Beniger 1986), we can try preliminary plots, not shown here, and then tweak the label positions away from their defaults (figure 15):

```
. generate pos = 3
. replace pos = 9 if inlist(year, 1900, 1920)
. replace pos = 10 if inlist(year, 1870, 1980)
. replace pos = 12 if inlist(year, 1880, 1970)
. triplot ag ind tert , c(1) max(100) mlabel(year) mlabsize(*0.7) clpat(solid)
> mlabvpos(pos)
```

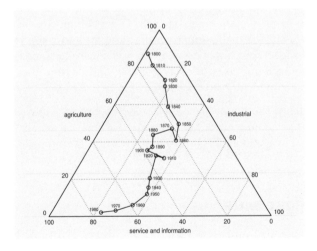

Figure 15: `triplot` is used here to show the changing composition of the U.S. civilian labor force from 1800 to 1980.

Again, we can reach through `triplot`, set up marker labels, and control their size and position, all by virtue of `triplot` being built upon `twoway`. In this case, the trajectory is fairly smooth from the highly agricultural economy of the early 19th century through a more industrial economy to nearer the present, dominated by services and information.

Superimposed on this smooth, almost evolutionary trend are some more complicated phases around the time of the Civil War and the First World War. Curiously, the Second World War has no discernible impact. The 10-year spacing of the series clearly filters out much detailed variation, but the graph retains interesting fine structure.

Very much complementary to this graph would be more standard line plots of the three sectors as time series. However, we can imagine adding a few other economies to the triangular plot without making it unreadable. The corresponding line plot with, say, a dozen time series would predictably be more difficult to read.

A major limitation on the usefulness of triangular plots as presented so far is that quite frequently data are crowded in a small part of the possible space. This is common with electoral data, especially if the third party or candidate attracts only a small percent of the total vote. Lumping all parties or candidates beyond first and second together for the sake of simplicity rarely solves this. Examples familiar to most readers are elections in the United States, long dominated by Democrats and Republicans, although often complicated in crucial ways by other parties or candidates. A variety of transformations have been suggested for triangular plots for this or other reasons. The simplest is a transformation suggested by Upton (2001). Extending `triplot` to accommodate this transformation is under way.

8 Conclusions

Many Stata users with categorical and compositional data tend to reach towards its tabulation routines in first examining datasets. This column has sampled only a few graphical possibilities in this field, some familiar staples and some possibly unfamiliar novelties, all of which are often more useful than is sometimes appreciated. Yet other graphical types have been proposed (Friendly 2000), and further Stata implementations in this territory may confidently be expected.

In the next column, the major theme will be comparison. How do we compare two or more subsets or variables, given various expectations, of equality, of additive and multiplicative shifts, and so forth? Once more, a variety of graphical types, tricks, and tips will be on display.

9 Acknowledgments

Elizabeth Allred, Ronán Conroy, Bob Fitzgerald, Roger Harbord, Friedrich Huebler, David Schwappach, Martyn Sherriff, Vince Wiggins, and Fred Wolfe made helpful comments during development of some programs discussed here. David Clayton, Anthony Edwards, and Graham Upton provided interesting discussions of triangular plots, which are likely to bear more fruit in future work on `triplot`.

10 References

Agresti, A. 2002. *Categorical Data Analysis.* 2nd ed. Hoboken, NJ: Wiley.

Aitchison, J. 1986. *The Statistical Analysis of Compositional Data.* London: Chapman & Hall/CRC.

Aitkin, M., D. Anderson, B. Francis, and J. Hinde. 1989. *Statistical Modelling in GLIM.* Oxford: Oxford University Press.

Atkinson, A. C. 1985. *Plots, Transformations, and Regression: An Introduction to Graphical Methods of Diagnostic Regression Analysis.* Oxford: Oxford University Press.

Beniger, J. R. 1986. *The Control Revolution: Technological and Economic Origins of the Information Society.* Cambridge, MA: Harvard University Press.

Bentley, J. L. 1988. *More Programming Pearls: Confessions of a Coder.* Reading, MA: Addison–Wesley.

Bentley, M. J., D. G. Hodgson, J. A. Smith, and N. J. Cox. 2005. Relative sea level curves for the South Shetland Islands and Marguerite Bay, Antarctic Peninsula. *Quaternary Science Reviews* 24: 1203–1216.

Blasius, J., and M. Greenacre, eds. 1998. *Visualization of Categorical Data.* San Diego: Academic Press.

Bross, I. D. J. 1958. How to use ridit analysis. *Biometrics* 14: 18–38.

Cleveland, W. S. 1993. *Visualizing Data.* Summit, NJ: Hobart.

———. 1994. *The Elements of Graphing Data.* Rev. ed. Summit, NJ: Hobart.

Cox, D. R., and E. J. Snell. 1989. *Analysis of Binary Data.* 2nd ed. London: Chapman & Hall.

Cox, N. J. 1999. gr41: Distribution function plots. *Stata Technical Bulletin* 51: 12–16. Reprinted in *Stata Technical Bulletin Reprints*, vol. 9, pp. 108–112. College Station, TX: Stata Press.

———. 2001. *Plotting graded data: A Tukey-ish approach.* Presentation to UK Stata Users Group meeting, Royal Statistical Society, London, 14–15 May. http://www.stata.com/support/meeting/7uk/cox1.pdf.

———. 2002. Speaking Stata: How to face lists with fortitude. *Stata Journal* 2: 202–222.

———. 2003a. Software update: gr41_1: Distribution function plots. *Stata Journal* 3: 211.

———. 2003b. Software update: gr41_2: Distribution function plots. *Stata Journal* 3: 449.

———. 2004. Speaking Stata: Graphing distributions. *Stata Journal* 4: 66–88.

Danuso, F. 1991. gr5: Triangle graphic for soil texture. *Stata Technical Bulletin* 2: 9–10. Reprinted in *Stata Technical Bulletin Reprints*, vol. 1, pp. 40–41. College Station, TX: Stata Press.

Diaconis, P. 1988. *Group Representations in Probability and Statistics*. Hayward, CA: Institute of Mathematical Statistics.

Duncan, O. D., H. Schuman, and B. Duncan. 1973. *Social Change in a Metropolitan Community*. New York: Sage.

Emerson, J. D. 1991. Introduction to transformation. In *Fundamentals of Exploratory Analysis of Variance*, ed. D. C. Hoaglin, F. Mosteller, and J. W. Tukey, 365–400. New York: Wiley.

Fienberg, S. E. 1980. *The Analysis of Cross-classified Categorical Data*. Cambridge, MA: MIT Press.

Fisher, R. A. 1940. The precision of discriminant functions. *Annals of Eugenics* 10: 422–429.

Fleiss, J. L., B. Levin, and M. C. Paik. 2003. *Statistical Methods for Rates and Proportions*. Hoboken, NJ: Wiley.

Flora, J. D. 1988. Ridit analysis. In *Encyclopedia of Statistical Sciences*, ed. S. Kotz and N. L. Johnson, vol. 8, 136–139. New York: Wiley.

Friendly, M. 2000. *Visualizing Categorical Data*. Cary, NC: SAS Institute.

Gower, J. C., and D. J. Hand. 1996. *Biplots*. London: Chapman & Hall.

Gray, J. 1993. Möbius's geometrical mechanics. In *Möbius and his Band: Mathematics and Astronomy in Nineteenth-century Germany*, ed. J. Fauvel, R. Flood, and R. Wilson, 79–103. Oxford: Oxford University Press.

Haberman, S. J. 1996. *Advanced Statistics Volume I: Description of Populations*. New York: Springer.

Howarth, R. J. 1996. Sources for a history of the ternary diagram. *British Journal for the History of Science* 29: 337–356.

Knoke, D., and P. J. Burke. 1980. *Log-linear Models*. Beverly Hills, CA: Sage.

Lloyd, C. J. 1999. *Statistical Analysis of Categorical Data*. New York: Wiley.

Marden, J. I. 1995. *Analyzing and Modeling Rank Data*. London: Chapman & Hall.

McCullagh, P., and J. A. Nelder. 1989. *Generalized Linear Models*. 2nd ed. London: Chapman & Hall/CRC.

Mitchell, M. 2004. *A Visual Guide to Stata Graphics*. College Station, TX: Stata Press.

Möbius, A. F. 1827. *Der barycentrische Calcul: ein neues Hülfsmittel zur analytischen Behandlung der Geometrie dargestellt und insbesondere auf die Bildung neuer Classen von Aufgaben und die Entwicklung mehrerer Eigenschaften der Kegelschnitte.* Leipzig: Johann Ambrosius Barth.

Parzen, E. 1993. Change *PP* plot and continuous sample quantile function. *Communications in Statistics—Theory and Methods* 22: 3287–3304.

Royston, P. 1996. gr21: Flexible axis scaling. *Stata Technical Bulletin* 34: 9–10. Reprinted in *Stata Technical Bulletin Reprints*, vol. 6, pp. 34–36. College Station, TX: Stata Press.

Simonoff, J. S. 2003. *Analyzing Categorical Data.* New York: Springer.

Stevens, S. S. 1946. On the theory of scales of measurement. *Science* 103: 677–680.

Stouffer, S. A., A. A. Lumsdaine, M. H. Lumsdaine, R. M. Williams, M. B. Smith, I. L. Janis, S. A. Star, and L. S. Cottrell. 1949. *The American Soldier: Combat and its Aftermath.* Princeton, NJ: Princeton University Press.

Tocher, J. F. 1908. Pigmentation survey of school children in Scotland. *Biometrika* 6: 129–235.

Townsend, J. G. 1995. *Women's Voices from the Rainforest.* London: Routledge.

Tukey, J. W. 1960. The practical relationship between the common transformations of percentages or fractions and of amounts. Reprinted in *The Collected Works of John W. Tukey. Volume VI: More Mathematical*, 1990, ed. C. L. Mallows, 211–219. Pacific Grove, CA: Wadsworth and Brooks/Cole.

———. 1961. Data analysis and behavioral science or learning to bear the quantitative man's burden by shunning badmandments. Reprinted in *The Collected Works of John W. Tukey. Volume III: Philosophy and Principles of Data Analysis: 1949–1964*, 1986, ed. L. V. Jones, 187–389. Monterey, CA: Wadsworth and Brooks/Cole.

———. 1977. *Exploratory Data Analysis.* Reading, MA: Addison–Wesley.

Upton, G. J. G. 2001. A toroidal scatter diagram for ternary variables. *American Statistician* 55: 247–250.

Wild, C. J., and G. A. F. Seber. 2000. *Chance Encounters: A First Course in Data Analysis and Inference.* New York: Wiley.

About the author

Nicholas Cox is a statistically minded geographer at the University of Durham. He contributes talks, postings, FAQs, and programs to the Stata user community. He has also co-authored fourteen commands in official Stata. He was an author of several inserts in the *Stata Technical Bulletin* and is Executive Editor of the *Stata Journal.*

The Stata Journal (2004)
4, Number 3, pp. 329–349

Speaking Stata: Graphing agreement and disagreement

Nicholas J. Cox
University of Durham, UK
n.j.cox@durham.ac.uk

Abstract. Many statistical problems involve comparison and, in particular, the assessment of agreement or disagreement between data measured on identical scales. Some commonly used plots are often ineffective in assessing the fine structure of such data, especially scatterplots of highly correlated variables and plots of values measured "before" and "after" using tilted line segments. Valuable alternatives are available using horizontal reference patterns, changes plotted as parallel lines, and parallel coordinates plots. The quantities of interest (usually differences on some scale) should be shown as directly as possible, and the responses of given individuals should be identified as easily as possible.

Keywords: gr0005, graphics, comparison, agreement, paired data, panel data, scatterplot, difference-mean plot, Bland–Altman plot, parallel lines plot, parallel coordinates plot, pairplot, parplot, linkplot, Tukey

1   Introduction

In one way or another, much statistical data analysis can be regarded as comparative. Exploratory projects often center on comparison of groups or of variables. Modeling projects can be formulated as comparisons of one or more model predictions with data on response variables. Are levels, spreads, shapes, trends, predictions, estimates, or whatever similar despite differences or different despite similarities? Above all, how should we best compare those features quantitatively in ways that summarize the data well but also direct attention to any important details? Broadly interpreted, the subject of comparison is wide enough to cover a large fraction of statistical science.

Graphics clearly have a major role to play in comparison. Already in this sequence of columns, we have seen Stata graphics commands used to compare distributions and categorical arrays (Cox 2004b,a). Other examples come readily to hand. The scatterplot, perhaps the most versatile weapon in the statistical graphics armory, is a device for comparing values of two variables, both individually and collectively. Indeed, little that is useful in statistical graphics does not afford comparisons.

In this column, we focus more narrowly. The key question of comparison discussed is assessing agreement or disagreement between two or more datasets or subsets with variables measured on the same scale. We will look at some official and user-written graphical programs available in Stata 8 for such problems. The emphasis is on making use of all the information in the data. Almost always, we need data reduction and, thus, seek a concise and simplified summary in terms of a few key parameters, quite possibly in terms of a formal model. Almost always, we need to consider the risk of losing valuable information by producing that summary. Graphics can guide our imagination, suggesting the best kind of summary. Graphics can also keep us honest, showing us the inadequacies of any particular summary.

2   Graphs as answers to questions

Let us start with a platitude and build on it. A platitude gets easy assent but can seem obvious and uninteresting. The challenge is to use that platitude as a platform for a discussion. An idea on which everyone can agree provides a starting point for analysis. Far from there being nothing to discuss fruitfully, there is then everything to discuss fruitfully.

The platitude is that a good graph is a good answer to a good question. That thought can be developed in various ways.

One way to develop it is to look at graphs and to try to determine the questions that they best answer. Sometimes it is then not clear which questions are being answered. Sometimes it is then not clear that the graph types being used are the most efficient forms for answering the questions being asked. Doing this even in a particular project is, unsurprisingly, often easier said than done. While a common prejudice runs that graphs are essentially trivial, a really good graph can prove extraordinarily elusive. Just as with photography and film, the most spectacular results may require almost endless experimentation and a ruthlessly critical attitude.

Another way to develop that thought is to look at questions and to try to identify the kinds of graphs that are most useful for different kinds of questions. This seems yet more difficult to do in a systematic way, exposing the lack of a good general theory of statistical graphics. The best available texts portray diverse repertoires of different kinds of graphs. The best available software is flexible enough to do most of what you can imagine. However, neither provides a theory.

As a starting point, questions may be classified as general:

What are these data like?

What patterns, trends, similarities, differences are there?

What interesting, informative, puzzling detail can we identify?

Or they may be specific:

Are these variables or subsets identical (as a reference case)?

Is the difference constant (e.g., 0), or is the ratio constant (e.g., 1)?

Has a transformation done what we wanted? Do the data now show a more tractable structure?

When answering general questions, we want graphs, above all, to provide summary and exposure (Tukey and Wilk 1966) and to show coarse and fine structure. The best general graphs allow both. At least in principle, dot plots and scatterplots show all the data. In practice, identical points will be overlaid on a scatterplot, and many similar points may be difficult to distinguish on either kind of plot. Even in principle, box plots and histograms do not show all the data. Reduction to selected quantiles and bin counts deliberately discards much detail, which quite possibly may be unintelligible noise, but also quite possibly be fine structure that we should be thinking about.

When answering specific questions, we want graphs to answer those questions directly without posing too many challenges. Any graph that requires extensive decoding or any other kind of hard work (say, rotating it mentally) is likely to be ineffective in practice.

3   Paired data

Let us now make this more concrete, starting with one very common structure in statistical science, that of paired data. Pairs arise, naturally or experimentally, in many circumstances: before and after, import and export, consumption and production, supply and demand, systolic and diastolic, left and right, husband and wife, control and treatment. In such situations, we often ask specific questions, such as whether means or, more generally, distributions can be considered identical. At worst, students—and often more experienced researchers too—go straight into a t test or a correlation without scrutinizing the data as a whole.

In Stata, paired data may be held in both wide and long data structures in the terminology of [D] **reshape**. For example, Dale et al. (1987) examined data on plasma β-endorphin concentrations in picomole/liter as measured in 11 runners before and after a "fun run" half-marathon. Endorphins are naturally occurring chemicals that can give relief from pain and make runners and indeed other people feel good through exercise. A wide structure accommodates such data in two variables, say `before` and `after`. In practice, an identifier variable will usually also appear.

```
. list
```

	id	before	after
1.	1	4.3	29.6
2.	2	4.6	25.1
3.	3	5.2	15.5
4.	4	5.2	29.6
5.	5	6.6	24.1
6.	6	7.2	37.8
7.	7	8.4	20.2
8.	8	9	21.9
9.	9	10.4	14.2
10.	10	14	34.6
11.	11	17.8	46.2

A long structure can be obtained from this by using `reshape`. The detail to note here is that some renaming of variables may be required first.

```
. rename before conc1
. rename after conc2
. reshape long conc, i(id) j(time)
(output omitted)
. list
```

	id	time	conc
1.	1	1	4.3
2.	1	2	29.6
3.	2	1	4.6
4.	2	2	25.1
5.	3	1	5.2
6.	3	2	15.5
7.	4	1	5.2
8.	4	2	29.6
9.	5	1	6.6
10.	5	2	24.1
11.	6	1	7.2
12.	6	2	37.8
13.	7	1	8.4
14.	7	2	20.2
15.	8	1	9
16.	8	2	21.9
17.	9	1	10.4
18.	9	2	14.2
19.	10	1	14
20.	10	2	34.6
21.	11	1	17.8
22.	11	2	46.2

The long structure may look more awkward here, but it is also natural if these data are considered as panel or longitudinal data, although with just two time points.

Given two paired variables, the most obvious graph that preserves the pairing is a simple scatterplot. In contrast, quantile–quantile or side-by-side dot plots, box plots, histograms, etc., lose the information on pairing. With scatterplots, however, it is easy to forget the specific question being asked. Much of our experience with scatterplots is based on checking scatters for linear relationships. But linearity $y = a + bx$ is more general than equality $y = x$, constant difference $y = a + x$, or constant ratio $y = bx$. If our idea—or our ideal—of underlying structure is not linearity in general but some special case, then a good graph will be one constructed so that ideal can be tested easily. (From a different but complementary perspective, Miller [1986, chapter 6] stresses how the models $Y = \Delta + X$ and $Y = \rho X$ are stopping points on the road to $Y = \alpha + \beta X$.)

One common practice is to superimpose a pertinent reference line on the scatterplot, most commonly $y = x$. In Stata 8, one idiom is

```
. scatter yvar xvar || function y = x, ra(xvar)
```

Longtime users may prefer something more like

```
. scatter yvar xvar xvar, ms(oh none) connect(none l) sort
```

itself an echo of an idiom common in Stata 7 and earlier:

```
. graph yvar xvar xvar, s(oi) c(.l) sort
```

A detail easy to overlook in the first way of doing it is `ra(`*xvar*`)`, without which the range defaults to (0,1). That default could be exactly what is wanted, or it could be such a minute fraction of the desired range that the line of equality is barely visible on the plot. It is also worth being aware of cosmetic options, such as `clpattern()` for varying the line pattern. Clearly, other reference lines may also be used, depending on what seems appropriate.

Such scatterplots with lines of equality superimposed are readily understood and often regarded as standard in various fields. But how efficient are they at answering the question? A problem with graphs with sloping reference lines is that it can be difficult to read the quantities of most interest. This leads immediately to the realization that other forms of graph are needed. We will look at two solutions suggested for this.

4 Horizontal reference lines

One golden rule is that horizontal alignment of reference lines makes comparisons much easier. It is very easy to check for patterns when the ideal plots as a flat configuration and very easy to check for departures from patterns when those departures are measured vertically. The eye and brain then have fewer challenges to overcome.

In mainstream statistical literature, this idea was repeatedly emphasized by John Tukey from at least the 1960s. He put it very well in his text *Exploratory Data Analysis* (1977, 148):

> Choosing scales to make behavior roughly linear always allows us to see local or idiosyncratic behavior much more clearly. Subtracting incomplete descriptions to make behavior roughly flat always allows us to expand the vertical scale and look harder at almost any kind of remaining behavior.

In other literatures, this idea has been rediscovered intermittently, for example, in medical statistics by Oldham (1962, 1968); Altman and Bland (1983); and Bland and Altman (1986, 1995a,b, 1999). No doubt other references can be supplied; please email the author if you know of any from the 1960s or earlier. In essence, it is one of the main ideas behind several kinds of residual plot (e.g., plotting residual versus fitted, residual versus predictor or other variable, etc.). However, even after at least 40 years of multiple reinvention, the principle still seems to deserve emphasis.

Thus, if differences are of interest, we could usefully plot differences $y - x$ versus means $(y+x)/2$. (The terminology Bland–Altman plot is common in medical statistics. Difference versus sum, a variant sometimes met, is clearly just the same plot apart from axis labeling.) If ratios are of interest, we could usefully plot ratios y/x versus geometric means $\sqrt{yx}$. Note that geometric means will be close to arithmetic means whenever y is close to x, so long as all values remain positive. For some groups of users, means will be familiar, but geometric means will not be, pushing the choice towards means.

The two cases of differences and ratios cover a large fraction of comparisons in practice. The Stata manipulations to produce graphs based on differences, means, ratios, and geometric means are easy. A few `generate` statements are all that is required. However, if you are doing this repeatedly, being able to do it on the fly is likely to be more convenient. One wrapper for this purpose is `pairplot` from SSC. You can install it by using the `ssc` command ([R] **ssc**) in an up-to-date, net-aware Stata:

```
. ssc install pairplot
```

`pairplot` is a wrapper program for `twoway`. Among its options are `diff`, `mean`, `ratio`, and `gmean`.

Let us look at an example in some detail. Glaciers gain mass mainly by accumulation of snow, especially in their upper parts, and lose mass by ablation, including melting and other means, especially in their lower parts. Notionally, there is an equilibrium-line altitude (elevation above sea level) for each glacier at which accumulation balances ablation. Monitoring this as it varies is of interest not just to glaciologists. Many other scientists are interested in such changes in time, space, or both as measures of glaciers' response to climatic and other fluctuations. In polar and mountain areas, direct and long-sustained meteorological records tend to be very sparse, so good proxies for climate are welcome.

Equilibrium line altitude (ELA) is best established by detailed field measurement on each glacier. The methods are simple in principle, using stakes placed in the ice, but repeated access to glacier surfaces can be difficult, expensive, and even dangerous, and there are too many glaciers for this to be anything other than exceptional. So there is great interest in proxies for the proxy, especially those that can be derived by map measurements (Leonard and Fountain 2002; Cogley and McIntyre 2003). The simplest proxy is (minimum glacier altitude + maximum glacier altitude)/2, often called the mean altitude, although many statistical people would want to call it a midrange. Another proxy is the contour (line of equal altitude) that is nearly straight, given that areas of accumulation tend to have concave contours and areas of ablation tend to have convex contours. One name for this is the kinematic ELA. There are protocols for complicated cases, and these ideas do not so readily apply to glaciers that end in water or on cliffs.

Leonard and Fountain (2002) collected data for 40 glaciers, mostly in the northern hemisphere, for which all three methods have been used. Correlations are generally very high:

```
. describe observed kinematic midrange

              storage  display    value
variable name  type    format     label    variable label

observed       int     %8.0g               mean observed ELA, m
kinematic      int     %8.0g               kinematic ELA, m
midrange       int     %8.0g               (min + max altitude)/2, m
. summarize observed kinematic midrange

    Variable |      Obs        Mean    Std. Dev.       Min        Max

    observed |       40     2820.55       1219.8        331       4934
   kinematic |       40    2675.125     1152.285        320       4840
    midrange |       40    2796.325       1225.8        300       4785
. correlate observed kinematic midrange
(obs=40)

             | observed kinema~c midrange

    observed |   1.0000
   kinematic |   0.9933    1.0000
    midrange |   0.9970    0.9897    1.0000
```

The associated scatterplots are correspondingly impressive (figure 1):

```
. graph matrix observed kinematic midrange
```

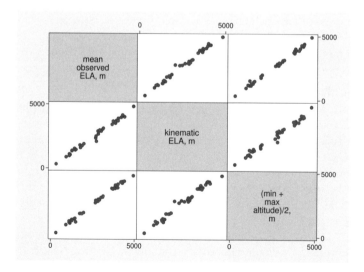

Figure 1: Scatterplots of three different altitude measures show very high correlations but fail to allow effective scrutiny of the fine structure of the data.

However, major reservations are in order. As Leonard and Fountain (2002) emphasize, the correlations are dominated by the large variations in altitudes from glacier to glacier and do not give a direct summary of the virtues of the different measurement methods. In any case, it is arguable that a more relevant single-number summary is the concordance correlation (Krippendorff 1970; Lin 1989, 2000; Steichen and Cox 2002, and references therein), which summarizes agreement (is y equal to x?), not linearity (is y equal to some $a + bx$?). But no single-number summaries can possibly do justice to any fine structure here, any more than a map can do the work of a microscope.

A `pairplot` of difference versus mean for `observed` and `kinematic` (figure 2) shows much more about the structure of errors than is evident from the scatterplot. The largest error is more than 600 m, and there seems to be some correlation between difference (`observed − kinematic`) and mean:

```
. pairplot observed kinematic, diff mean yla(, ang(h))
```

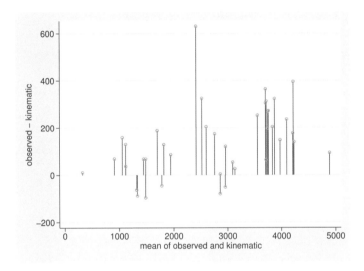

Figure 2: A plot of difference versus mean for observed and kinematic ELA shows more about the structure of the errors.

The correlation between difference and mean is 0.442 (the sign is arbitrary, as it depends on which way the difference is calculated).

Incidentally, this correlation is a test statistic for a null hypothesis of equal variances given bivariate normality (Pitman 1939; also see Snedecor and Cochran 1989, 192–193). We set that consideration on one side and prefer to use it as an exploratory diagnostic. The ideal is clear: difference and mean should be uncorrelated. Nevertheless, if you seek tests, as well as measures, in this terrain, you might also be interested in an F test of equality of means and variances, again assuming bivariate normality, proposed by Bradley and Blackwood (1989). The prospect of investigating how these tests perform when bivariate normality breaks down will appeal or appall, according to statistical taste.

Back to the example: as it turns out, an offset between observed ELA and kinematic ELA is expected from the physics of glacier flow (Leonard and Fountain 2002). A resistant estimate of that offset is the median of the differences, 135 m, and this can be used as a base for the vertical spikes (figure 3):

```
. pairplot observed kinematic, diff mean base(135) t1(base 135 m, place(w))
> yla(, ang(h))
```

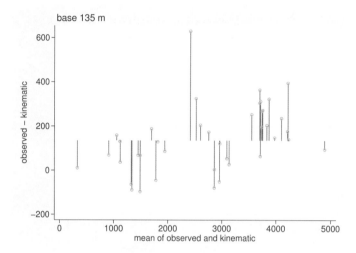

Figure 3: Difference spikes are expressed relative to a base of 135 m, the median of the differences.

observed functions here as a "gold standard" measurement, so some might prefer to show that directly on the graph. The rule with **pairplot** is that a third variable is used for the other axis (figure 4). (The option **mean** is correspondingly not specified.)

```
. pairplot observed kinematic observed, diff base(135) t1(base 135 m, place(w))
> yla(, ang(h))
```

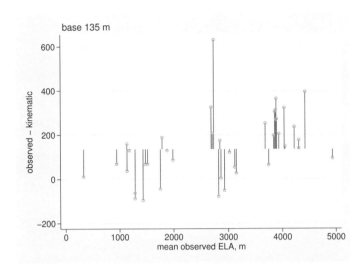

Figure 4: Observed ELA is shown on the horizontal axis, given its role as a gold standard.

We should flag here that Bland and Altman (1995b) warn explicitly that you should *not* plot difference against standard. See this paper immediately as an antidote to the example just set. At the same time, note that the graph just given does resemble that of difference versus mean.

The apparent tilt could be handled empirically by regression, but it would be better to have a physical understanding of why it occurred. In addition, any regression would have to look the problem of errors in variables squarely in the eye (see, for example, Dunn 2004). In any case, `midrange` behaves much better (figure 5):

```
. pairplot observed midrange observed, diff yla(, ang(h))
```

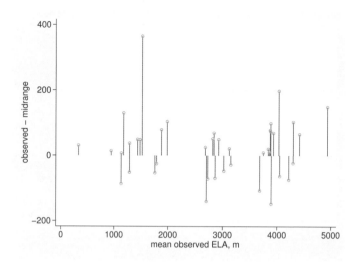

Figure 5: A plot of difference versus mean for observed ELA and midrange altitude shows a more attractive error structure.

The maximum and median error (23 m) are much smaller here, and no tilt is apparent. The correlation between difference and mean—that is, $(\text{observed} + \text{midrange})/2$— is much smaller at 0.063; again, the sign is at choice. Leonard and Fountain (2002) prefer `kinematic` to `midrange`, a preference that is not well supported by their results and is certainly not supported by those here. As seen above, on average, `kinematic` has a mean 145 m below that of `observed`, while `midrange` has a mean 24 m below, which results resemble the differences of 130 m and 35 m reported by Cogley and McIntyre (2003). The smaller offset, the lack of tilt and, indeed, the simpler measurement method point to `midrange` as the better proxy in this dataset.

5   Parallel line plots

Evidently, being able to plot difference-mean and ratio-geometric mean plots was one motivation for writing `pairplot`. Another was the excellent paper by McNeil (1992), who focuses particularly on what is most effective for before and after comparisons of the kind particularly common in medical statistics. His main example is the beta-endorphin dataset given earlier.

A popular plot for showing such data is as tilted line segments, as shown in figure 6. (Later we will see how to get such plots in Stata.)

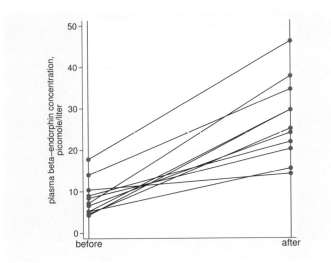

Figure 6: Change in endorphin concentration is shown by tilted line segments.

One rationale for such a plot could be that "before" and "after" define a time direction, as is explicit in a long data structure. Drawing a time axis thus appears natural. Certainly, more data on how concentrations varied in time for these runners would make a time-series graph the most obvious form. But in the special case of just two times of observation, how effective is this kind of plot? In Cleveland's (1994) terms, change has been encoded essentially as slope of segment and must be decoded from that representation. That is, the segment slopes are visually the most obvious element—the reader has to work much harder to estimate differences on this scale.

Two other simple and very practical limitations of this kind of plot need underlining. As emphasized by McNeil (1992), even in a graph for a sample of eleven the criss-crossing of lines is confusing to decode. That problem is naturally accentuated with much larger sample sizes. A related problem is that it is difficult to relate the line segments to identifiers. As given to us, the identifiers just indicate sort order and are otherwise not informative. In a typical research project, however, investigators might well try to interpret anomalous individuals using other information, whether qualitative or quantitative. Hence, easy relation to identifiers is highly desirable.

Such considerations led McNeil (1992) to suggest representing individuals by parallel lines. In this kind of plot, values are coded by positions along a common scale and changes by lengths of line segments, direct and effective choices, typically much easier to decode mentally. Vertical parallel lines are the default of pairplot. As before, a third variable, when supplied, is plotted on the other axis (figure 7):

```
. pairplot before after id, xla(1/11, labsize(medium))
> yti("plasma beta-endorphin concentration," "picomole/liter") aspect(1)
> yla(, ang(h))
```

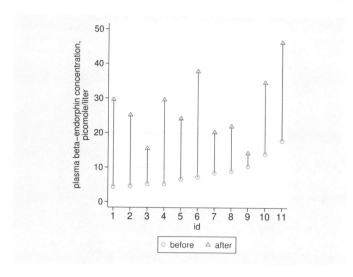

Figure 7: Change in endorphin concentration is shown by vertical line segments, with a third variable on the other axis.

Although with this dataset, a call with just two variables would have produced an almost identical plot, as the *x*-axis defaults to observation number (figure 8):

```
. pairplot before after, xla(1/11, labsize(medium))
> yti("plasma beta-endorphin concentration," "picomole/liter") aspect(1)
> yla(, ang(h))
```

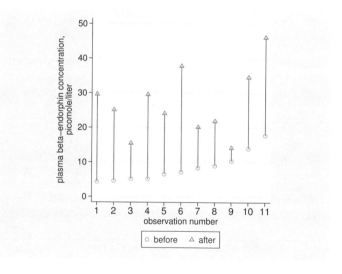

Figure 8: Change in endorphin concentration is shown by vertical line segments, with observation number on the other axis: in this case, the graph is almost identical.

Horizontal parallel lines merely require a `horizontal` option (figure 9):

```
. pairplot before after id, yla(1/11, ang(h) labsize(medium)) horiz
> xti("plasma beta-endorphin concentration," "picomole/liter")
> yti(, orient(horiz)) aspect(1) yscale(reverse)
```

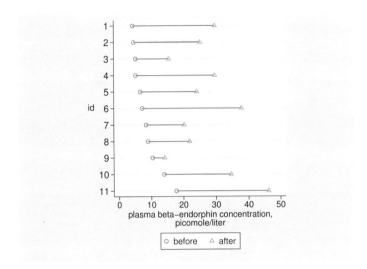

Figure 9: Change in endorphin concentration is shown by horizontal line segments, with a third variable on the other axis.

You may recognize that this last graph is just a short step away from those produced by default by `graph dot`. The similarity is made more evident by adding the option `blstyle(none)`. Conversely, although it appears to be undocumented, `graph dot` supports vertical alignment given a `vertical` option. `graph dot` can show more than two variables if required, but there are no options to emphasize the line segments between values. `pairplot` is, as the name implies, limited to comparisons between two variables, but it is intended to be flexible for that problem.

An aside on these data: McNeil (1992) uses a logarithmic scale for all graphs from the dataset. Revisiting the example, he uses the original scale (McNeil 1996, 52–54). Most important here is how the changes behave. To a good approximation, change in concentration is independent of original, so a logarithmic transformation is neither needed nor helpful. At first sight, when we look at these data, logarithms may seem a more natural scale: concentrations tend to have skewed distributions, and more importantly, they might be expected to change multiplicatively rather than additively. How the data behave is nevertheless the crucial issue.

6   pairplot: a reprise

A synopsis of `pairplot` may be useful here. This is not a complete summary. As usual, the help file gives other details.

`pairplot` supports plots with

Y1. two variables, linked vertically, on the y-axis or

Y2. their difference, shown vertically, on the y-axis or

Y3. their ratio, shown vertically, on the y-axis and

X1. order of observations on the x-axis or

X2. a specified variable on the x-axis or

X3. sort order on some *varlist* (ascending or descending) on the x-axis or

X4. mean of two variables on the x-axis or

X5. geometric mean of two variables on the x-axis or

any of the above combinations but with axes reversed.

7   Beyond pairs to parallel coordinates plots

The case of paired data is common, interesting, and important. We should now look at more general comparisons. As before, the focus is on graphs that show visual linkage of specific individuals (however defined) across two or more variables or groups. It is easy enough to juxtapose or superimpose several distributions (as histograms, dot plots, quantile functions, distribution functions, etc.), but that is not the aim here.

One standard device is now usually known as a parallel coordinates plot. You may well know the idea under a less formal name, for example, as a profile plot. Wegman (1966) gives an accessible, lucid, and definitive account for a statistical readership. He nevertheless understates the long history and wide geography of such plots: the nineteenth century French railway schedules beloved of Tufte (1983) show the main idea, as do plots used for many decades to show the results of so-called "bumps" rowing races, held at Oxford and Cambridge in Britain and occasionally elsewhere. Any time-series plot is just a twist away from a parallel coordinates plot. At most, one idea is needed, that of interchanging of variables and observations.

That restructuring, on the fly, is the main need within any Stata implementation of parallel coordinates plots. An implementation for Stata 4 was given by Gleason (1996). An implementation for Stata 8 is the `parplot` program downloadable from SSC. (The name is close to `pairplot`, but no confusion between the two has been evident. A name `parcoord` would invite confusion with Gleason's program, and a name `parcplot` is otherwise objectionable.)

An example is worth a thousand words, and we have already had one. The origin of figure 6 may now be revealed:

```
. parplot before after, tr(raw)  yla(, ang(h))
> yti("plasma beta-endorphin concentration," "picomole/liter")  aspect(1)
> xla(, labsize(medium))
```

`parplot` is a wrapper for `twoway connected`. The options in this example are thus standard `twoway` options, with the exception of `tr(raw)`, that is, `transform(raw)`, which specifies showing data on the original or raw scale. The default of `parplot` is to scale values on each variable to (value − minimum)/(maximum − minimum), a choice that permits comparison of variables measured in quite different units or with quite different scales. Being able to cope with such data is crucial for many applications of `parplot`, although it is beyond the main theme here. Other transforms are also possible, and logarithmic scales are available as usual through `xscale()` or `yscale()`. Against the semantic objection that a raw transform is no transform must be set the fact that StataCorp has already laid claim to the option name `scale()`.

A more substantial example can be seen by putting together the three glacier altitude variables (figure 10).

```
. parplot kinematic observed midrange, tr(raw) aspect(1)
> xla(1 `""kinematic" "ELA""´ 2 `""mean observed" "ELA""´
>     3 `""midrange" "altitude""´)
> yla(, ang(h)) ytitle(metres, orient(horiz))
```

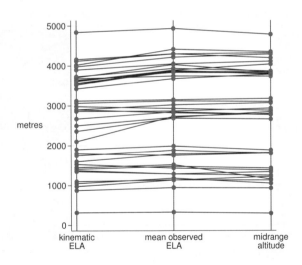

Figure 10: Three measures of altitude are compared by a parallel coordinates plot.

As before, the syntax is mostly standard, and only a few details need comment. Environmental scientists usually find it congenial to plot altitude on a vertical axis, even though it is not the response variable. Note also that the reference case of equal values for the three variables defines horizontal profiles, departures from which are easy to see. In other circumstances, as might be guessed, a `horizontal` option exchanges axes as compared with the default.

There is space enough to change the aspect ratio to square. A parallel coordinates plot is map as well as microscope but retains enough fine structure—indeed, in a strong sense, all the information in the data—to underline visually that midrange altitude is closer quantitatively to mean observed ELA than is kinematic ELA.

A different example uses United States census data from 1980. Marriages, divorces, and deaths come as absolute numbers, so calculating rates relative to population and taking logarithms wrench the data towards comparable, indeed similar, scales. Different transforms of the data may be tried, but the default `maxmin` scale works well (figure 11), as does a horizontal alignment:

```
. sysuse census, clear
(1980 Census data by state)

. foreach v in death divorce marriage {
  2.          generate r_`v´ = log10(`v´ / pop)
  3. }

. parplot r_*, horiz by(region, caption(logarithmic scales)
> title(US states 1980)) yla(1 "deaths" 2 "divorces" 3 "marriages", ang(h))
```

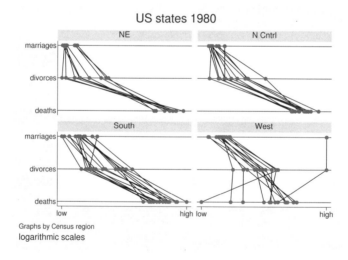

Figure 11: Three demographic measures are compared for U.S. states within census regions by a parallel coordinates plot.

Here, and elsewhere, there is room for thought and experiment about not only the scale or transformation to be used but also the order of the variables. Putting closely correlated variables close together and negating variables to make as many positive correlations as possible are two of the possible tricks for making an effective graph. At the same time, there is little point in including variables that are very poorly correlated with any others, unless that is precisely the point you want to make.

Substantively, this example shows nothing that is not well known about the demographic characteristics of the United States, nor was that the intent. The heterogeneity of the "West" again should come as no surprise. In terms of method, however, parallel coordinates plots have uses when you are seeking cluster structure or checking how far cluster structure exists. For example, if a few distinct groups do exist in the data, then this should be evident in a parallel coordinates plot. Conversely, indications that the data form a continuum rather than a set of clusters might lead the investigator to call off a pointless cluster analysis.

8 A timely comment

Time-series people, and many others, may think that some obvious points have been overlooked so far, so we need to underline one connection hinted at already. With a long data structure and both panel and time variables, the endorphin data can be declared as panel data. Then we can use dedicated time-series plots, such as `xtline` (added to Stata on 12 September 2003; thus, it is not documented in Stata 8 manuals) (figure 12):

```
. tsset id t
        panel variable:  id, 1 to 11
         time variable:  time, 1 to 2
. xtline conc, overlay legend(off) xla(1 "before" 2 "after", labsize(medium))
> yla(, ang(h)) yti("plasma beta-endorphin concentration," "picomole/liter")
> aspect(1)
```

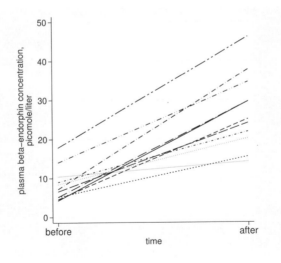

Figure 12: Treating the endorphin data as panel data allows `xtline` to be used for an alternative plot.

In this example, the options have been chosen to emphasize the similarity with earlier graphs, particularly figure 6. A difference not evident in the printed graph here is that `xtline` can make use of different pen colors for different panels, which may be attractive to you. The main graphical point made earlier was that, even for such data, tilted line segments are a relatively ineffective kind of display. Nevertheless, in terms of Stata possibilities, do note that if your data are in or near panel-data form, then you may find this path an easy one to take.

Yet another way to do it is provided by the `linkplot` program on SSC, which does not assume panel or even time-series data but is more general. You should find it easy to download `linkplot` using `ssc` and to explore its possibilities yourself.

9   Conclusion

Stata graphics takes one from the lows of struggling to get the details you desire to the highs of being able to think freely about the aim, form, and content of your displays. The richness of commands on offer is all based on the completely rewritten (and still evolving) Stata 8 graphics engine. In this column, we have looked at a classic and fundamental problem in statistical science, assessing agreement on a shared scale. Key to graphical success, both here and elsewhere, is linking the design of graphics to the questions of greatest importance and interest. Surprisingly often, commonly used plots (such as scatterplots of highly correlated variables or plots of values measured "before" and "after" using tilted line segments) can be ineffective because they fail to show the most pertinent quantities directly.

Those who have been following these columns will recognize how many of the ideas can be traced directly to John W. Tukey (1915–2000) or to his students and collaborators; see Brillinger (2002) and related papers for excellent recent appreciations of Tukey's work. In one classic paper, Tukey (1972, 293) distinguishes between various kinds of graphs, including propaganda graphs "intended to show the reader what has already been learnt" and analytical graphs "intended to let us see what may be happening over and above what we have already described". Arguably, there are too many propaganda graphs, even in statistical science. Fortunately, many forms of analytical graphs have been devised that are easy to understand and to use, such as those based on horizontal reference patterns, changes plotted as parallel lines, or parallel coordinates plots.

In the next issue, we will complete the quartet of graphics columns promised for 2004 by discussing plots for model diagnostics.

10   Acknowledgments

Thomas Steichen first told me about concordance correlation and intensified my interest in these problems. Ian Evans alerted me to the glacier altitude problem and the particular papers cited. Erik Beecroft, Bob Fitzgerald, and Vince Wiggins made helpful comments on programs discussed here. Ronán Conroy, Patrick Royston, and Anders Skrondal gave help with references.

11   References

Altman, D. G., and J. M. Bland. 1983. Measurement in medicine: The analysis of method comparison studies. *Journal of the Royal Statistical Society, Series D* 32: 307–317.

Bland, J. M., and D. G. Altman. 1986. Statistical methods for assessing agreement between two methods of clinical measurement. *Lancet* i: 307–310.

———. 1995a. Comparing two methods of clinical measurement: A personal history. *International Journal of Epidemiology* 24: S7–S14.

———. 1995b. Comparing methods of measurement: Why plotting difference against standard method is misleading. *Lancet* 346: 1085–1087.

———. 1999. Measuring agreement in method comparison studies. *Statistical Methods in Medical Research* 8: 135–160.

Bradley, E. L., and L. G. Blackwood. 1989. Comparing paired data: A simultaneous test for means and variances. *American Statistician* 43: 234–235.

Brillinger, D. R. 2002. John W. Tukey: his life and professional contributions. *Annals of Statistics* 30: 1535–1575.

Cleveland, W. S. 1994. *The Elements of Graphing Data*. Rev. ed. Summit, NJ: Hobart.

Cogley, J. G., and M. S. McIntyre. 2003. Hess altitudes and other morphological estimators of glacier equilibrium lines. *Arctic, Antarctic, and Alpine Research* 35: 482–488.

Cox, N. J. 2004a. Speaking Stata: Graphing categorical and compositional data. *Stata Journal* 4: 190–215.

———. 2004b. Speaking Stata: Graphing distributions. *Stata Journal* 4: 66–88.

Dale, G., J. A. Fleetwood, A. Weddell, and R. D. Ellis. 1987. Beta endorphin: A factor in "fun run" collapse. *British Medical Journal* 294: 1004.

Dunn, G. 2004. *Statistical Evaluation of Measurement Errors: Design and Analysis of Reliability Studies*. London: Arnold.

Gleason, J. R. 1996. gr18: Graphing high-dimensional data using parallel coordinates. *Stata Technical Bulletin* 29: 10–14. Reprinted in *Stata Technical Bulletin Reprints*, vol. 5, pp. 53–60. College Station, TX: Stata Press.

Krippendorff, K. 1970. Bivariate agreement coefficients for reliability of data. In *Sociological Methodology*, ed. E. F. Borgatta and G. W. Bohrnstedt, vol. 2, 139–150. San Francisco: Josssey-Bass.

Leonard, K. C., and A. G. Fountain. 2002. Map-based methods for estimating glacier equilibrium-line altitudes. *Journal of Glaciology* 49: 329–336.

Lin, L. I.-K. 1989. A concordance correlation coefficient to evaluate reproducibility. *Biometrics* 45: 255–268.

———. 2000. A note on the concordance correlation coefficient. *Biometrics* 56: 324–325.

McNeil, D. 1992. On graphing paired data. *American Statistician* 46: 307–311.

———. 1996. *Epidemiological Research Methods*. Chichester, UK: Wiley.

Miller, R. G. 1986. *Beyond ANOVA: Basics of Applied Statistics*. New York: Wiley. Reprinted, London: Chapman & Hall, 1997.

Oldham, P. D. 1962. A note on the analysis of repeated measurements of the same subjects. *Journal of Chronic Diseases* 15: 969–977.

———. 1968. *Measurement in Medicine: The Interpretation of Numerical Data*. London: English Universities Press.

Pitman, E. J. G. 1939. A note on normal correlation. *Biometrika* 31: 9–12.

Snedecor, G. W., and W. G. Cochran. 1989. *Statistical Methods*. 8th ed. Ames, IA: Iowa State University Press.

Steichen, T. J., and N. J. Cox. 2002. A note on the concordance correlation coefficient. *Stata Journal* 2: 183–189.

Tufte, E. R. 1983. *The Visual Display of Quantitative Information*. Cheshire, CT: Graphics Press.

Tukey, J. W. 1972. Some graphic and semigraphic displays. In *Statistical Papers in Honor of George W. Snedecor*, ed. T. A. Bancroft and S. A. Brown, 293–316. Ames, IA: Iowa State University Press.

———. 1977. *Exploratory Data Analysis*. Reading, MA: Addison–Wesley.

Tukey, J. W., and M. B. Wilk. 1966. Data analysis and statistics: An expository overview. *AFIPS Conference Proceedings* 29: 695–709.

Wegman, E. J. 1966. Data analysis and statistics: An expository overview. *Journal of the American Statistical Association* 85: 664–675.

About the author

Nicholas Cox is a statistically minded geographer at the University of Durham. He contributes talks, postings, FAQs, and programs to the Stata user community. He has also co-authored fourteen commands in official Stata. He was an author of several inserts in the *Stata Technical Bulletin* and is Executive Editor of the *Stata Journal*.

The Stata Journal (2004)
4, Number 4, pp. 449–475

Speaking Stata: Graphing model diagnostics

Nicholas J. Cox
University of Durham, UK
n.j.cox@durham.ac.uk

Abstract. Plotting diagnostic information calculated from residuals and fitted values is a long-standard method for assessing models and seeking ways of improving them. This column focuses on the statistical mainstream defined by regression models for continuous responses, treated in a broad sense to include (for example) generalized linear models. After some comments on the history of such ideas (and even their anthropology and psychology), the commands available in official Stata are reviewed, and a `modeldiag` package is introduced. A detailed example on fuelwood yield from fallow areas in Nigeria illustrates a variety of general points and specific tips.

Keywords: gr0009, modeldiag, anovaplot, indexplot, ofrtplot, ovfplot, qfrplot, racplot, rdplot, regplot, rhetplot, rvfplot2, rvlrplot, rvpplot2, graphics, diagnostics, regression, generalized linear models, analysis of variance

1   Introduction

A common task in statistical graphics is looking at various flavors of residual and predicted (fitted) values after fitting a model. There are now many ideas on how these extra values may be used graphically to examine the fit between data and models and to seek possible means of improving models.

Diagnostic graphs have a key role in adding fine structure to judgments based, all too often, largely on single-valued summaries, such as R^2 (whether plain, adjusted, or pseudo-), AIC, or BIC. Naturally, these graphs should also complement, in a heuristic or exploratory manner, inferences based on specific tests of hypotheses.

Several different kinds of graph may be inspected in many modeling exercises, partly because each kind may be best for particular purposes and partly because in many projects a variety of models—in terms of functional form, choice of predictors, and so forth—may be entertained, at least briefly. It is therefore helpful to be able to produce such graphs very rapidly.

The focus here is on what may be fairly regarded as a central part of statistical modeling: regression treated in a suitably broad sense but emphasizing the modeling of continuous response variables. Thus we will not recapitulate problems or tools specific to particular areas, such as survival or time-series analysis, or material on categorical responses, covered so thoroughly with reference to Stata by Long and Freese (2003).

After a swift survey of the history of these ideas and some comments on variations in current practice, we will review official Stata commands and the `modeldiag` package. A detailed example closes the column.

2   History, anthropology, and psychology

As with many statistical ideas, the approaches discussed in this paper have both very long and much shorter roots. The idea of a residual, as a difference between observed and expected, is some centuries old and is exemplified in the experimental work of Galileo (who also had the idea that error distributions were likely to be symmetric and unimodal). For Galileo's statistical attitudes in particular and other related ideas in the 17th and 18th centuries, see Hald (1986) and Plackett (1988). The general method of inference from residual phenomena (meaning appearances) was strongly emphasized by John Herschel (1792–1871) in *A Preliminary Discourse on the Study of Natural Philosophy* (1830). His book has been described as the first work on philosophy of science in English written by a working scientist; it was widely influential in the 19th century, being studied carefully by Charles Darwin, among many others (Ruse 1979). Apposite quotations from Herschel's book appear as chapter epigraphs in the statistical monograph of Cook and Weisberg (1982).

Despite this splendid past, reports on analysis of residuals and the use of graphs to look at the results of regression-like models both appear to have been unusual in the literature before the early 1960s. (Scatterplots of raw data were naturally more common.) Before modern computers, and also afterwards, the usual algorithms for regression and analysis of variance led to sums of squares, mean squares, and the associated test statistics, rather than the set of individual residuals. In the face of calculation work that could be very time-consuming, just calculating a set of residuals may often have seemed a complication too far, even when data analysts thought about it. In addition, easy production of presentable graphs was available only to rather few researchers until very recently. Even many statistical packages produced ugly lineprinter graphs until the early 1990s. On the other hand, it is difficult to assess how often good data analysts looked at tables or even graphs of residuals informally before it became respectable, and even fashionable, to do so and to talk about it in print.

One striking exception to the general dearth before about 1960 of residual analysis appears in the work of the Danish scientist Thorvald Nicolai Thiele (1838–1910). Thiele worked in astronomy, mathematics, actuarial science, and statistics. He advocated graphical analysis of residuals checking for trends, symmetry of distributions, and changes of sign, and even warned against over-interpreting such graphs (Thiele 1889; Lauritzen 2002, 180–182).

Whatever the detailed prehistory, the modern history of such diagnostic graphics can be said to begin in the early 1960s with the work of Frank Anscombe, John W. Tukey, and others. Major references include Anscombe (1961), Tukey (1962), and Anscombe and Tukey (1963). (Incidentally, Anscombe [1918–2001] and Tukey [1915–2000] married sisters, which led Tukey to refer to Anscombe as his brother-in-squared-law.) Ideas percolated into textbooks, for example, Draper and Smith (1966, 1981, 1998)—their third edition remains a friendly and quite comprehensive survey of regression. The approach was soon defined by its own monographs (Belsley, Kuh, and Welsch 1980; Cook and Weisberg 1982; Atkinson 1985). The field is still active, with many new ideas that cannot be explored here (Cook 1998; Atkinson and Riani 2000).

Nevertheless this forty-year period has evidently been too short to establish any strong uniformity of methodology. Forays into intellectual anthropology or psychology appear necessary to explain some marked variations in practices from field to field. The logic of fitting and assessing regression-like models should transcend disciplinary boundaries, but fields do vary in what is preferred, or even compulsory, showing contrasts in tribal habits. There are fields like my own (geography, environmental sciences) in which a strongly graphical approach is not only welcome but positively expected. There are also fields in which regression-like modeling is central but use of diagnostic graphics appears rare and almost all the emphasis in model assessment is on figures of merit and formal test statistics.

Informal conversations bring up two points repeatedly to explain disinclinations to adopt a strongly graphical approach. First, researchers who may be working with a large number of variables often feel that there would just be too many graphs to work with, especially if many of those graphs could be equivocal or contradictory in their indications. The number of predictors can indeed be limiting, but there are also useful general graphs that are possible regardless of that number. Second, and seemingly more crucial, is that analysis practices tend to be dominated by whatever formats journals prefer or require for publication of results. Frequently, ritual displays of coefficients, standard errors, confidence intervals, and the like are considered essential but requests for graphical displays would be regarded as idiosyncratic or as posing unreasonable requests for journal space.

More detailed discussion of tribal habits in the use of regression would take us too far afield. For an incisive and much broader critique of many issues in contemporary regression methodology, see the polemical monograph of Berk (2004).

3 Existing commands in official Stata

Those with experience in using Stata for both modeling and graphics will often find it easy to get diagnostic graphs with just a few command lines. Thus suppose that you are checking an assumption that error terms follow a normal or Gaussian distribution. The best way to do this graphically is usually with a probability (meaning quantile–quantile) plot with ordered residuals on one axis and the corresponding expected quantiles on the other axis. Arguably that is much more informative than either a general purpose test (chi-square, Kolmogorov–Smirnov) or even a specific purpose test (Shapiro–Wilk, Shapiro–Francia). (The latter two are, in effect, producing numerical summaries of the information in the probability plots.)

In Stata, once a model has been run, this process is at most two commands: a `predict` command to get the residuals and a `qnorm` command to get the graph. Often the residuals will have been calculated already for another purpose, so only one command is needed. In this case, no special command appears needed, but if you were doing this repeatedly, the saving from two lines to one could make it worth your while to put the commands into a short wrapper program. (A related graph showing two quantile plots side by side will be discussed in more detail later.)

Conversely, I often produce observed versus fitted plots, on which more will also be said later. This also requires a `predict` to get the fitted values and a `scatter` to get the graph. At this point, however, I usually want to add a reference line of equality, and I realize that I want better axis titles. The last requires some labeling of the fitted variable or an axis title specification. I found myself doing this so frequently that an `ovfplot` with sensible defaults became a practical proposition.

Official Stata supplies a built-in bundle of commands originally written for use after `regress` and thus *post hoc* in character:

> `avplot` and `avplots`

> `cprplot` and `acprplot`

> `lvr2plot`

> `rvfplot` and `rvpplot`

These were introduced in Stata 3.0 in 1992 and are documented at [R] **regression diagnostics**. More recently, in an update to Stata 7.0 in 2001, all but the first two were modified so that they may be used after `anova`.

Despite their many uses, this suite omits some very useful kinds of plots, while none of the commands may be used after other modeling commands. To make that point concrete, I find the logic behind generalized linear models very compelling and often want to use `glm`. I also find that physically inspired models typically lead to the brute force approach of `nl`. Evidently, none of the standard commands just mentioned will work after either of these.

Different in spirit, but worth a strong recommendation, are a bundle of commands introduced as part of the new graphics of Stata 8. `twoway lfit`, `twoway qfit`, `twoway fpfit`, and their kin implement models on the fly for various functional forms, namely linear, quadratic, and fractional polynomials. They give graphs of data and fitted curves and (if desired) confidence intervals. In this territory, note also commands such as `lowess` and `locpoly` (see Gutierrez, Linhart, and Pitblado 2003).

Clearly, users have a choice. They can explore data using these latter commands and follow up graphs that appear successful with the formalities of, e.g., `regress` or `fracpoly`. I find it particularly helpful whenever an informal (or perhaps semiformal) exploration with `lowess` or `locpoly` either supports the notion of a linear approximation or convinces me that I need something quite different. These explorations rarely survive to the printed page, but they can nevertheless be invaluable aids in model development.

Alternatively, users may find that a simple model developed using such modeling commands can be plotted using one of these `twoway` types. So `twoway lfit` may also be used *post hoc* whenever a `regress` with one predictor appears adequate or at least interesting. It matters not if the next idea is then that curvature or some other nonlinearity or some obvious outliers need more care and attention.

Equally clearly, these `twoway` commands are limited to a few simple and standard forms and in no sense exhaust the repertoire of models that might be useful.

In this column, the main story concerns the use of a new set of commands, which as implied are biased to graphics useful for models predicting continuous response variables. The ideal followed in producing this set is to make minimal assumptions about which modeling command has been issued previously. The downside for users is that if the data and the previous model results do not match the assumptions, it is possible to get either bizarre results or an error message, but these are constitutional hazards in any case. More positively, it is the prerogative, and also the responsibility, of the user to decide what is justifiable. The programs discussed here are available with the Stata Journal software. In addition, they may be downloaded from SSC; see [R] **ssc** for details.

4 The modeldiag package

The principles followed in programming such commands include

- as far as possible, the command name by itself should produce a useful plot

- `predict` is used to produce temporary variables for residuals, fitted values, etc.

- each graph refers to the last model fitted

- each graph has reasonably smart default axis titles, etc.

- graphs implement Stata 8 graphics

- options are provided for key needs

The commands which have been written are as follows. First comes a group of general-purpose commands.

4.1 ovfplot

`ovfplot` plots observed versus fitted values for the response from an immediately previous `regress` or similar command, with a line of equality superimposed by default. Some merits of this kind of plot deserve mention. It is easy to understand, especially for presentation to users who do not specialize in statistical applications, and indeed it is often among many scientists' lists of favorites. It is generally applicable, as many kinds of models lead to predictions directly comparable with the response variable, whatever the number of predictors or the functional form. Residuals, measured on the same scale as the response, can be read off the plot as vertical differences. Another merit, which is also a profound defect, is that the observed versus fitted plot can be an optimistic or propaganda plot (Tukey 1972; Cox 2004b), making even a lousy model look fairly good.

4.2   regplot

`regplot` plots fitted or predicted values from an immediately previous `regress` or similar command. By default, the data for the response are also plotted.

With one syntax, no variable name is specified: `regplot` then shows the response and predicted values on the y-axis and the predictor named first in the `regress` or similar command on the x-axis. Thus with this syntax the plot shown is sensitive to the order in which predictors are specified in the estimation command.

With another syntax, a variable name is supplied, which may name any numeric variable: this is then used as the variable on the x-axis.

Thus in practice, `regplot` is most useful when the fitted values are a smooth function of the variable shown on the x-axis, or a set of such functions given also one or more dummy variables as predictors. However, other applications also arise, such as plotting observed and predicted values from a time-series model versus time.

By default, `regplot` shows the fitted values using `twoway mspline`. The `plottype()` option may be used to specify another `twoway` plottype.

A `separate()` option specifies that values of fitted and observed responses be plotted as separate groups corresponding to the distinct values of the variable specified. This is especially useful when a categorical predictor has been included in the model as one or more dummy variables. The `by()` option remains available as usual.

Note that `regplot` does not work after `anova`; see the comments on `anovaplot`, discussed later.

4.3   rvfplot2

`rvfplot2` plots residuals versus fitted values from an immediately previous `regress` or similar command. This is one of the main workhorses in this area. The underlying idea is that no news is good news: ideally, the scatter should be fairly even and patternless, with no hints of (for example) curvature, uneven scatter, or disturbance by outliers.

The residuals are, by default, those calculated by `predict, residuals` or (if the previous estimation command was `glm`) by `predict, response`. The fitted values are those produced by `predict` by default after each estimation command. `rvfplot2` is offered as a generalization of `rvfplot` in official Stata.

There is support for specifying several types of residual other than the default. An `rscale()` option specifies a transformed scale on which to show the residuals using Stata syntax and `X` as a placeholder for the residual variable name. Thus `rscale(X^2)` specifies squaring, to show relative contribution to residual variance; `rscale(abs(X))` specifies absolute value, to set aside sign; `rscale(sqrt(abs(X)))` specifies root of absolute value, a useful scale on which to check for heteroskedasticity.

Similarly, an `fscale()` specifies a transformed scale on which to show the fitted values using Stata syntax and `X` as a placeholder for the fitted variable name. Thus, for example, `fscale(2 * ln(X))` specifies twice the natural logarithm, which is the constant information scale for a generalized linear model with gamma error. Similarly, arguments of `2 * sqrt(X)`, `2 * asin(sqrt(X))`, and `-2 / sqrt(X)` specify the constant information scale for Poisson, binomial, and inverse Gaussian errors, respectively. See McCullagh and Nelder (1989, 398) for background.

A `lowess` option specifies that the residuals will be smoothed as a function of the fitted using `lowess` (options of which may be specified in turn).

4.4   rvpplot2

`rvpplot2` plots residuals versus values of a specified predictor (a.k.a., independent variable or carrier) from an immediately previous `regress` or similar command. The residuals are, by default, those calculated by `predict, residuals` or (if the previous estimation command was `glm`) by `predict, response`.

`rvpplot2` is offered as a generalization of `rvpplot` in official Stata.

There is support for specifying several types of residuals other than the default. A `force` option allows you to specify a predictor variable not included in the previous model. An `rscale()` option specifies a transformed scale on which to show the residuals using Stata syntax and `X` as a placeholder for the residual variable name. Thus `rscale(X^2)` specifies squaring, to show relative contribution to residual variance; `rscale(abs(X))` specifies absolute value, to set aside sign; `rscale(sqrt(abs(X)))` specifies root of absolute value, a useful scale on which to check for heteroskedasticity. A `lowess` option specifies that the residuals will be smoothed as a function of the predictor using `lowess` (options of which may be specified in turn).

Thus `rvpplot2` offers scope for easy plotting against either a predictor already in the model or a possible predictor not at present included in the model. The latter could be time or spatial position if there was concern about serial autocorrelation. A clear pattern in this plot will point up the possibility of modifying the model. With a predictor in the model, some evidence of curvature or other nonlinearity might point to a change in how it was included, for example, the addition of a quadratic term or a prior transformation. With a predictor not in the model, evidence of correlation might suggest adding that predictor to the model.

4.5   indexplot

`indexplot` plots estimation results (by default whatever `predict` produces by default) from an immediately previous `regress` or similar command versus observation number (i.e., `_n`).

Values are shown, by default, as vertical spikes starting at 0 using `twoway dropline`, but the graph may be recast to another `twoway` plot type.

4.6   qfrplot

`qfrplot` plots quantiles of fitted values, minus their mean, and quantiles of residuals from the previous `regress` or similar command.

Fitted values are whatever `predict` produces by default, and residuals are whatever `predict, res` produces. Comparing the distributions gives an overview of their variability and some idea of their fine structure, as plots appear side by side with aligned vertical scales. Note that the rationale of this graph is comparing distributions of (fitted − mean) and residuals; hence, it is vital that residuals be measured on the same scale as the response.

Options include observed versus normal (Gaussian) quantile–quantile plots. Note that `qfrplot` is essentially a wrapper for calls to `qplot` (Cox 2004a,c), itself a generalization, apart from one small detail, of official Stata's `quantile` command.

Cleveland (1993) gives many side-by-side quantile plots of fit and residuals, which he calls "residual-fit spread plots". See, for example, the graph on his page 41. However, he also uses this term for side-by-side time-series plots of fit and residuals (page 157). The command name and description here emphasize the use of a quantile plot.

4.7   rdplot

`rdplot` plots residual distributions from the previous `regress` or similar command. The residuals are, by default, those calculated by `predict, residuals` or (if the previous estimation command was `glm`) by `predict, response`.

The graph by default is a single or multiple dotplot, as produced by `dotplot`. Histograms as produced by `histogram` or box plots as produced by `graph box` or `graph hbox` may be selected. Oneway plots as implemented in `onewayplot`, skewness plots as implemented in `skewplot`, or quantile plots as implemented in `qplot` may also be selected. On the last three, see Cox (2004a,c).

Various options offer scope for grouping residuals in various ways, according to values of some other variable (e.g., a predictor).

4.8   rhetplot

`rhetplot` checks for residual heteroskedasticity after the previous `regress` or similar command.

`rhetplot` graphs standard deviations (optionally variances) of residuals for distinct groups formed by combinations of specified variables; standard deviations (optionally variances) of residuals against means of groups of a specified variable; or standard deviations (optionally variances) of residuals against means of groups of fitted values.

The residuals are, by default, those calculated by `predict, residuals` or (if the previous estimation command was `glm`) by `predict, response`. There is support for specifying several types of residual other than the default.

The graph is produced by `lowess`. The "smooth" curve shown (unless the number of groups specified is very small) is best regarded as an informal indication of the general pattern of variability of residuals.

Next come a group of commands designed for models based on time, although they may easily be extended to other situations when appropriate. Even commands that assume a previous `tsset` can be applied after `sorting` to a sensible order, and

```
. generate t = _n
. tsset t
```

Conversely, the safeguard of requiring `tsset` gives users some protection against getting incorrect results, in particular with panel data.

4.9   ofrtplot

`ofrtplot` plots observed, fitted, and residuals versus "time" variables after the previous `regress` or similar command. It is primarily designed for time-series models, and by default the predictor is whatever has been `tsset` as the time variable. However, other variables may be specified, whether or not data have been `tsset`.

Observed values are for the response or dependent variable from the last model, fitted values are whatever `predict` produces by default, and residuals are whatever `predict, res` produces.

By default, the plot has two panels. In the top panel, observed and fitted are plotted against the predictor. In the bottom panel, residuals are plotted against the predictor, by default as spikes from zero. Optionally, plots may be superimposed, not separate.

4.10   rvlrplot

`rvlrplot` plots residuals versus lagged (i.e., lag 1) residuals for time-series data after the previous `regress` or similar command. Data must have been `tsset` previously.

By default, residuals are whatever `predict, res` produces after a model. There is support for specifying several types of residuals other than the default.

4.11   racplot

`racplot` plots the residual autocorrelation function after the previous `regress` or similar command. `racplot` calculates the residuals and then fires up `ac`. Data must have been `tsset` previously. There is support for specifying several types of residuals other than the default. An `rscale()` option specifies a transformed scale for the residuals using Stata syntax and `X` as a placeholder for the residual variable name.

Finally, in this listing, is a command especially dedicated to the results of `anova`:

4.12 anovaplot

`anovaplot` plots fitted or predicted values from an immediately previous one-, two-, or three-way `anova`. By default, the data for the response are also plotted. In particular, `anovaplot` can show interaction plots. The format of the graph may be varied by permuting the names of predictors used in `anova`.

Note especially that the graph format produced by `anovaplot` is appropriate for models with at most one continuous predictor, which should always be the predictor named first. With that caveat, `anovaplot` offers a way of showing parallel and diverging regression lines for models with one continuous predictor.

It is curious that analysis-of-variance people typically draw interaction plots but suppress the data, whereas regression people prefer to draw scatterplots showing both observed and fitted values. Admittedly, a complicated set of crossing lines showing interactions may seem to leave little scope for showing data effectively, while a relatively simple regression leaves plenty of scope, but the difference is nevertheless intriguing.

5 Example: wood volumes and fallow length in Nigeria

In many areas of the humid tropics, fallow areas are used for fuelwood and are indeed vitally important for local energy supply. With increasing population pressure and intensification of cropping, such fallows are under threat. The growth of trees on fallows is thus of great interest. The data for this example come from a paper by Adesina (1990), who looked at 80 fallows near Gbongan township in western Nigeria, asking: do fallows planted with the fast-growing species *Gliricidia sepium* yield more wood than "natural" or self-propagated fallows?

Data were collected for quadrats of 20 m × 20 m on slopes no greater than 2°, 40 on each of two fallow types. Of several variables measured, we will look at wood volume, in cubic meters, as a response, and length of fallow in years and fallow type, coded by 0 for natural and 1 for *Gliricidia*, as predictors.

The main purpose of this section is to provide some simple illustrations of `modeldiag` in action, without purporting to give an analysis fully sensitive to all the scientific or practical nuances of the problem. Adesina's paper gives a most interesting description of the context but an analysis that includes no graphs and is based on bivariate correlations and *t* tests comparing means. Here we seize the opportunity to use length of fallow fully as a quantitative predictor within various models.

```
. scatter volume years, by(type) ms(oh)
```

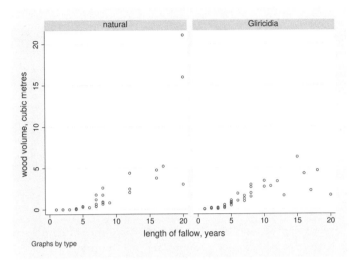

Figure 1: Scatterplot. The relationship between wood volume, fallow length, and vegetation type is obscured by the skewness of both volume and length.

This exploratory scatterplot (figure 1) and some simple summary statistics immediately reveal various basic features. Both `volume` and `years` are positively skewed (the moment measure of skewness is 4.372 in the case of `volume`), and relatively few plots have been fallow for more than (say) 10 years (17/80). It is thus difficult to discern structure, especially on the right-hand side of the plot. At worst, some outliers may be present. It will be easier to see what is going on if we transform the response. Cube root, sometimes a rather arbitrary transform, seems very natural here, as the units then become meters, and we are dealing with what may be called an equivalent length. Imagine a bundle of wood with given volume and of cubical shape; its sides will have this length. (In passing, note an excellent article on dimensional analysis and statistics by Finney [1977].) The cube root of volume is still skewed (0.691), but we are moving in the right direction and not making anything worse.

```
. generate curtvol = volume^(1/3)
. label variable curtvol "equivalent length, m"
. regress curtvol years
```

Source	SS	df	MS
Model	18.4151275	1	18.4151275
Residual	5.33350034	78	.068378209
Total	23.7486278	79	.300615542

```
Number of obs =      80
F(  1,    78) =  269.31
Prob > F      =  0.0000
R-squared     =  0.7754
Adj R-squared =  0.7725
Root MSE      =  .26149
```

curtvol	Coef.	Std. Err.	t	P>|t|	[95% Conf.	Interval]
years	.0919356	.0056022	16.41	0.000	.0807826	.1030886
_cons	.2591998	.0494769	5.24	0.000	.1606989	.3577007

More importantly, a trial regression looks good numerically, with clear-cut F and t results, R^2 of 0.775, and root mean squared error of 0.261 m.

However, a basic `regplot` shows that the *Gliricidia* data points possess considerable curvature, so that the model is missing some structure (figure 2). This is also shown by the residual versus fitted plot (figure 3).

```
. regplot, by(type)
```

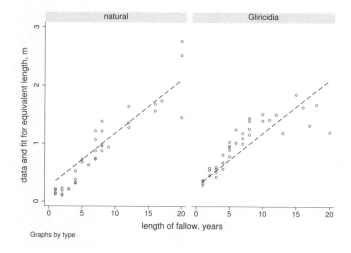

Figure 2: Regression plot. The regression of equivalent length on fallow length still leaves important curvature, especially for *Gliricidia*.

```
. rvfplot2, ms(oh)
```

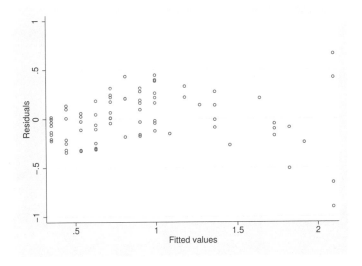

Figure 3: Residual versus fitted plot. Curvature is evident here too.

We add `type` as a dummy variable and the product of `type` and `years` as another predictor to permit differing slopes and intercepts. This boosts R^2 to 0.830 and reduces root mean squared error to 0.230 m. Unsurprisingly, much curvature remains (figures 4, 5, and 6). Note that in this case—with just one predictor whose coefficient is positive—the residual versus fitted plot and the residual versus predictor plot are the same graph, modulo the labeling of the x-axis. Nevertheless one may be more convenient to read than the other, depending on whether the scientist finds it easier to think on the predictor or the response scale.

```
. generate type_years = type * years
. regress curtvol years type type_years
```

Source	SS	df	MS			
Model	19.7122058	3	6.57073528			
Residual	4.03642201	76	.053110816			
Total	23.7486278	79	.300615542			

Number of obs = 80
F(3, 76) = 123.72
Prob > F = 0.0000
R-squared = 0.8300
Adj R-squared = 0.8233
Root MSE = .23046

curtvol	Coef.	Std. Err.	t	P>|t|	[95% Conf. Interval]	
years	.1104499	.0067043	16.47	0.000	.0970971	.1238028
type	.4315831	.0873577	4.94	0.000	.257595	.6055711
type_years	-.0388651	.0099265	-3.92	0.000	-.0586355	-.0190947
_cons	.0446065	.0615559	0.72	0.471	-.0779928	.1672058

. regplot, by(type)

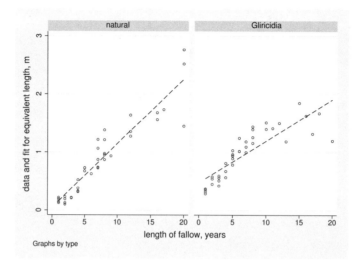

Figure 4: Regression plot. Allowing interaction is one thing, but the curvature for *Gliricidia* is another.

. rvfplot2, by(type) ms(oh)

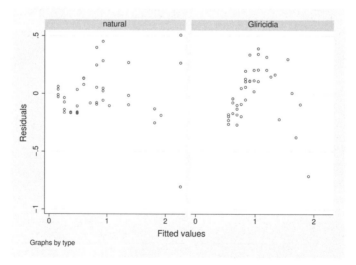

Figure 5: Residual versus fitted plot. Another way of seeing the curvature.

```
. rvpplot2 years, by(type) ms(oh)
```

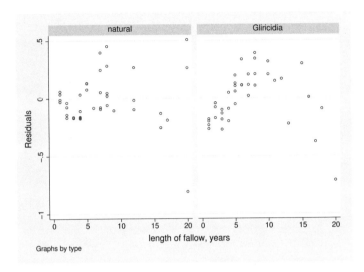

Figure 6: Residual versus predictor plot. Yet another way of seeing the curvature, in this case arguably clearer than the residual versus fitted plot.

A multiple dot plot shows that the residuals are heteroskedastic (figure 7). In using `rdplot`, there is a trade-off: enough groups are needed to get an idea of any fine structure, but not so many that there are too few data points in each group to summarize effectively. I often start with `group(3)` and go to more groups only if it seems sensible, but with a much larger dataset than that here, a larger number of groups would be justifiable.

```
. rdplot, group(3) ms(oh)
```

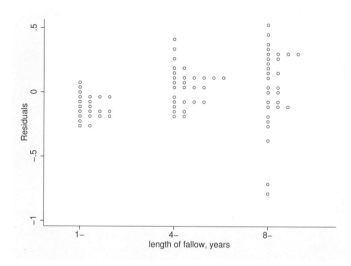

Figure 7: Residual distribution plot. Data points sliced into three groups according to values of the predictor with approximately the same size. Heteroskedasticity is evident.

So far then, although the `regress` output shows us a well-behaved dataset, the regression model is still failing to capture important structure. In addition, although it happens that the intercept is just above zero, there is no guarantee with a model of this kind that predictions are all positive, which is essential biologically. A positive prediction can be ensured by using a generalized linear model with logarithmic link. Using this, rather than a logarithmic transformation, has the signal advantage that results are returned on an intelligible scale, without any need for back-transformation. We have also some flexibility over choice of error distribution.

However, a logarithmic link with `years` as predictor would imply exponential growth over time, seemingly not appropriate for either type of vegetation. A power function appears more sensible than an exponential, which leads us to try log of years as a predictor.

```
. generate logyears = log(years)
. generate type_logyears = logyears * type
. glm curtvol logyears type type_logyears, link(log) nolog
```

```
Generalized linear models                    No. of obs       =         80
Optimization        : ML: Newton-Raphson     Residual df      =         76
                                             Scale parameter =  .0428134
Deviance            =  3.253821965           (1/df) Deviance =  .0428134
Pearson             =  3.253821965           (1/df) Pearson  =  .0428134

Variance function: V(u) = 1                  [Gaussian]
Link function     : g(u) = ln(u)             [Log]
Standard errors   : OIM

Log likelihood    =  14.57277095             AIC              = -.2643193
BIC               = -329.7802023
```

curtvol	Coef.	Std. Err.	z	P>|z|	[95% Conf. Interval]	
logyears	.9140198	.0595632	15.35	0.000	.7972781	1.030762
type	1.046955	.1861204	5.63	0.000	.6821658	1.411744
type_logye~s	-.4247248	.0752181	-5.65	0.000	-.5721496	-.2773
_cons	-1.947677	.1541104	-12.64	0.000	-2.249727	-1.645626

With appropriate extra predictors to allow for an interaction, the generalized linear model looks good numerically. The output for `glm` does not supply an R^2 or a root mean squared error, but a simple program not documented here summons up both these measures from the correlation and differences between response and fitted. For detailed statistical arguments on why doing this is perfectly sensible, see Zheng and Agresti (2000). The values of 0.864 and 0.207 m suggest some progress.

In looking at the model and data overall we can specify plotting against **years**, even though **logyears** was the predictor in the model (figure 8). Now the residuals look better (figures 9 and 10).

. regplot years, by(type)

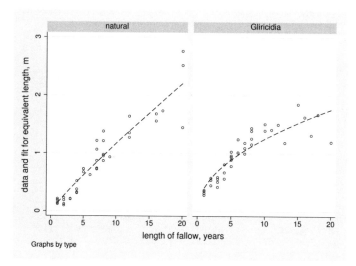

Figure 8: Regression plot. Generalized linear model fitted with logarithmic link and log years as one predictor, but plotted here versus years. This does a better job of capturing the different behavior.

. rvfplot2, by(type) ms(oh)

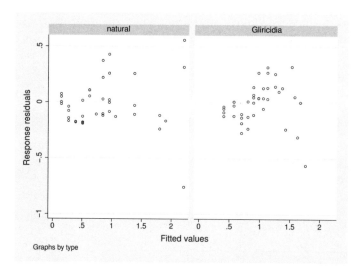

Figure 9: Residual versus fitted plot. The residuals are better behaved. Does important curvature persist?

. rvpplot2 years, by(type) ms(oh) force

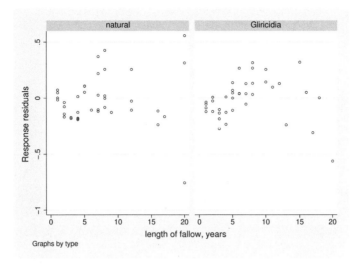

Figure 10: Residual versus predictor plot. Years is not in the model but can be used as an axis with the `force` option.

The quantile plot of fitted and residuals popularized by Cleveland (1993) is a nice summary of "how far we have come" compared with "how far we have yet to go" (figure 11). By default, we chose a normal (Gaussian) error family for the generalized linear model, so looking at residuals on a Gaussian scale is pertinent (figure 12). Nevertheless this plot avoids a key issue, whether error distributions are homoskedastic, which does not appear to be the case (figure 13). This leads to a switch to a gamma error family.

. qfrplot

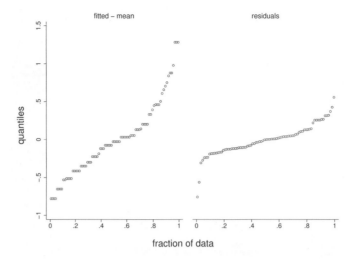

Figure 11: Quantile plot of fitted and residuals. Fitted − mean and residuals have the same scale, so they can be juxtaposed.

. qfrplot, gauss

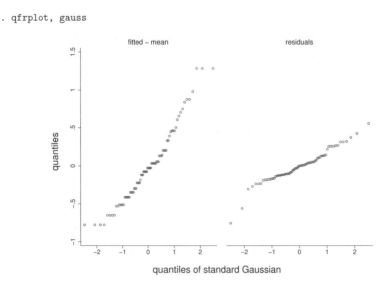

Figure 12: Quantile plot of fitted and residuals. A Gaussian scale is used, as that is the error family postulated. The assumption looks fair for the data as a whole.

```
. rdplot, group(3) ms(oh)
```

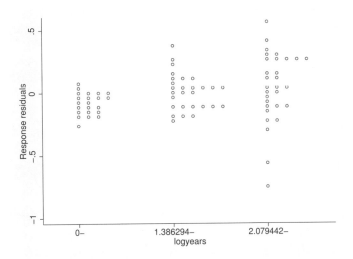

Figure 13: Residual distribution plot. The heteroskedasticity is still evident.

In one sense, the results for the fit look very similar, and every *P*-value in sight is excellent. R^2 drops a smidgen to 0.856 and root mean squared error rises correspondingly to 0.214 m, but experience teaches us not to be oversensitive to such small differences. The model gives a near-linear power function (power 0.964) for natural fallows and one much closer to the square root for *Gliricidia* ($0.964 - 0.383 = 0.581$) (figure 14). The residuals give no great cause for concern (figures 15 and 16); a sharp eye would wonder if the curvature of *Gliricidia* had been followed quite correctly, but there is an issue of how much weight to put on values for longer fallow lengths.

```
. glm curtvol logyears type type_logyears, link(log) f(gamma) nolog
Generalized linear models                      No. of obs        =         80
Optimization         : ML: Newton-Raphson      Residual df       =         76
                                               Scale parameter =  .0572479
Deviance         =   4.702906734               (1/df) Deviance =  .0618804
Pearson          =   4.350838306               (1/df) Pearson  =  .0572479
Variance function: V(u) = u^2                  [Gamma]
Link function    : g(u) = ln(u)                [Log]
Standard errors  : OIM
Log likelihood   = -56.88833941                AIC               =   1.522208
BIC              = -328.3311175
```

curtvol	Coef.	Std. Err.	z	P>|z|	[95% Conf. Interval]	
logyears	.9642254	.0430776	22.38	0.000	.8797948	1.048656
type	.9961841	.1186355	8.40	0.000	.7636627	1.228705
type_logye~s	-.3831086	.0638372	-6.00	0.000	-.5082273	-.2579899
_cons	-2.076803	.0821826	-25.27	0.000	-2.237878	-1.915728

```
. regplot years, by(type)
```

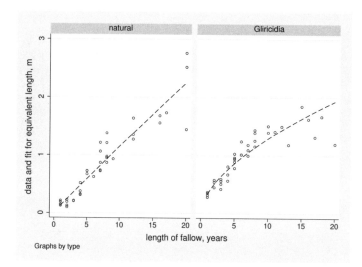

Figure 14: Regression plot. Fitted curves with gamma error assumption are similar to those with Gaussian error assumption.

```
. qfrplot
```

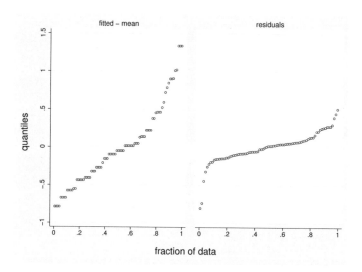

Figure 15: Quantile plot of fitted and residuals. A graphical alternative to an R^2 result.

```
. rdplot, group(3) ms(oh)
```

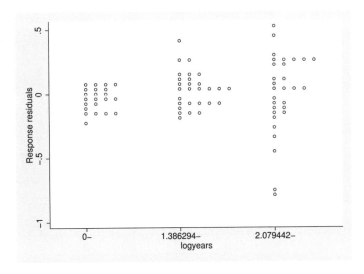

Figure 16: Residual plot. Heteroskedasticity is now expected given the gamma error assumption.

Offstage we saved the predictions from `glm` with Gaussian and gamma errors. After some prior surgery with `separate`, here is a line plot (figure 17). Both the similarity and the differences make sense. The gamma-based model is not constrained by an ideal of homoskedastic errors and is thus less sensitive to some rather low response values for longer fallow lengths, which seem rather suspicious. Until more ideas or more data arrive, the gamma-based model appears to have the edge.

```
. line p?a*0 p?a*1 years, sort yti("equivalent length, m")
```

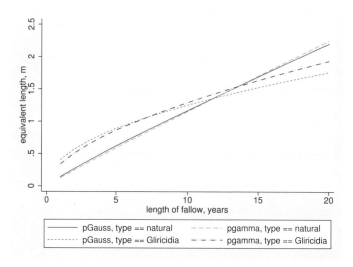

Figure 17: Line plot. The predictions for Gaussian and gamma error models are compared.

An observed versus fitted plot usually looks pretty good and provides an optimistic close (figure 18).

```
. ovfplot
```

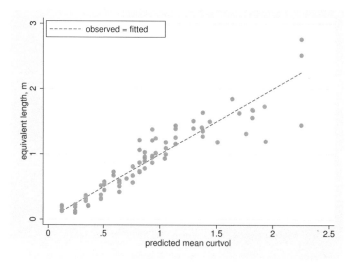

Figure 18: Observed versus fitted plot. A simple summary possible for many models.

As every modeler knows, fitting the data (or the apparent structure in the data) is only part of the battle. The ultracynical could comment that this example is, at root, showing that trees grow over time. Underlying the data, although at some removes from what we have, are presumably some monotonic growth curves. However, we have not data for individual trees monitored over time, but collective results for plots with varying number of trees from a spatial survey (cross-section, not panel data, in other words). No doubt we would benefit from extra predictors that might explain variations between plots, giving detail, say, on soils, microclimate, moisture, nutrients, and the precise history and pattern of land use at each site. However, these further predictors are not available and would have required an enormously bigger project. Adesina does give information on tree and herb diversity, which might serve as surrogates for competition from other species, but results not shown here indicate that they do not help substantially in improving the model, and in any case one should be wary of over-fitting. It is known that some wood is lost from fallows by casual harvesting before they are cleared for subsequent cultivation. Some of the data points for long fallows do look suspiciously low. One radical way to tackle this would be to repeat the analysis with only the shorter fallows, particularly as the comparison between natural and *Gliricidia* in early years is the heart of the matter.

6 Conclusions

The calculation of residuals—the bits left over, the parts of the data the model failed to reach—is the end of one process but also the beginning of a new one whenever we can see something that the model does not capture. Seeing that something is helped by having pictures to look at. We have shown that several Stata commands, official and user-written, exist to help.

What is exciting for Stata users interested in this approach is that many methods remain to be implemented in Stata, such as the ideas of Cook (1998) and Atkinson and Riani (2000). In addition, an open question is how far (and how) `avplot`, `avplots`, `cprplot`, `acprplot`, and `lvr2plot` may be generalized to be used with other commands. In this and other ways, the area of graphical diagnostics may be expected to develop further.

7 Acknowledgments

Kit Baum, Denis de Crombrugghe, Phil Ender, Ken Higbee, and Andy Sloggett commented on earlier versions of programs discussed here.

8 References

Adesina, F. A. 1990. Planted fallows for sustained fuelwood supply in the humid tropics. *Transactions, Institute of British Geographers* 15: 323–330.

Anscombe, F. J. 1961. Examination of residuals. *Proceedings of the Fourth Berkeley Symposium on Mathematical Statistics and Probability* 1: 1–36.

Anscombe, F. J., and J. W. Tukey. 1963. The examination and analysis of residuals. *Technometrics* 5: 141–160.

Atkinson, A. C. 1985. *Plots, Transformations, and Regression: An Introduction to Graphical Methods of Diagnostic Regression Analysis*. Oxford: Oxford University Press.

Atkinson, A. C., and M. Riani. 2000. *Robust Diagnostic Regression Analysis*. New York: Springer.

Belsley, D. A., E. Kuh, and R. E. Welsch. 1980. *Regression Diagnostics*. New York: Wiley.

Berk, R. A. 2004. *Regression Analysis: A Constructive Critique*. Thousand Oaks, CA: Sage.

Cleveland, W. S. 1993. *Visualizing Data*. Summit, NJ: Hobart.

Cook, R. D. 1998. *Regression Graphics: Ideas for Studying Regressions through Graphics*. New York: Wiley.

Cook, R. D., and S. Weisberg. 1982. *Residuals and Influence in Regression*. New York: Chapman & Hall.

Cox, N. J. 2004a. Software update: gr42_2: Quantile plots, generalized. *Stata Journal* 4: 97.

———. 2004b. Speaking Stata: Graphing agreement and disagreement. *Stata Journal* 4: 329–349.

———. 2004c. Speaking Stata: Graphing distributions. *Stata Journal* 4: 66–88.

Draper, N. R., and H. Smith. 1966. *Applied Regression Analysis*. New York: Wiley.

———. 1981. *Applied Regression Analysis*. 2nd ed. New York: Wiley.

———. 1998. *Applied Regression Analysis*. 3rd ed. New York: Wiley.

Finney, D. J. 1977. Dimensions of statistics. *Journal of the Royal Statistical Society, Series C* 26: 285–289.

Gutierrez, R. G., J. M. Linhart, and J. S. Pitblado. 2003. From the help desk: Local polynomial regression and Stata plugins. *Stata Journal* 3: 412–419.

Hald, A. 1986. Galileo's statistical analysis of astronomical observations. *International Statistical Review* 54: 211–220.

Herschel, J. F. W. 1830. *A Preliminary Discourse on the Study of Natural Philosophy.* London: Longman, Rees, Orme, Brown, and Green; John Taylor. Facsimile reprint: Chicago: University of Chicago Press, 1987.

Lauritzen, S. L. 2002. *Thiele: Pioneer in Statistics.* Oxford: Oxford University Press.

Long, J. S., and J. Freese. 2003. *Regression Models for Categorical Dependent Variables Using Stata.* Rev. ed. College Station, TX: Stata Press.

McCullagh, P., and J. A. Nelder. 1989. *Generalized Linear Models.* 2nd ed. London: Chapman & Hall/CRC.

Plackett, R. L. 1988. Data analysis before 1750. *International Statistical Review* 56: 181–195.

Ruse, M. 1979. *The Darwinian Revolution: Science Red in Tooth and Claw.* Chicago: University of Chicago Press.

Thiele, T. N. 1889. *Forlæsinger over Almindelig Iagttagelseslære: Sandsynlighedsregning og Mindste Kvadraters Methode.* Copenhagen: C. A. Reitzel. English translation included in Lauritzen 2002.

Tukey, J. W. 1962. The future of data analysis 33: 1–67 and 812.

———. 1972. Some graphic and semigraphic displays. In *Statistical Papers in Honor of George W. Snedecor*, ed. T. A. Bancroft and S. A. Brown, 293–316. Ames, IA: Iowa State University Press.

Zheng, B., and A. Agresti. 2000. Summarizing the predictive power of a generalized linear model. *Statistics in Medicine* 19: 1771–1781.

About the author

Nicholas Cox is a statistically minded geographer at the University of Durham. He contributes talks, postings, FAQs, and programs to the Stata user community. He has also co-authored fifteen commands in official Stata. He was an author of several inserts in the *Stata Technical Bulletin* and is Executive Editor of the *Stata Journal*.

Speaking Stata: Density probability plots

Nicholas J. Cox
Durham University, UK
n.j.cox@durham.ac.uk

Abstract. Density probability plots show two guesses at the density function of a continuous variable, given a data sample. The first guess is the density function of a specified distribution (e.g., normal, exponential, gamma, etc.) with appropriate parameter values plugged in. The second guess is the same density function evaluated at quantiles corresponding to plotting positions associated with the sample's order statistics. If the specified distribution fits well, the two guesses will be close. Such plots, suggested by Jones and Daly in 1995, are explained and discussed with examples from simulated and real data. Comparisons are made with histograms, kernel density estimation, and quantile–quantile plots.

Keywords: gr0012, density probability plots, distributions, histograms, kernel density estimation, quantile–quantile plots, statistical graphics

1 Introduction: how do methods become standard?

Many more methods are invented for analyzing data than ever become part of anybody's standard toolkit. What selection processes determine which methods become popular?

Optimists suppose that all really good ideas will sooner or later become noticed and accepted as sound and useful. Indeed, a good idea may be reinvented several times over: smart people will follow the same logical paths, albeit at different times and in varying circumstances. If Gauss had never got around to publishing least-squares, Legendre did anyway. For optimists, the selection of good methods is virtually Darwinian, but without the messy details of sex and death.

Pessimists suppose that the conservatism inherent in many aspects of teaching and research means that few data analysts look far beyond what they had learned in graduate school or what their peer group currently favors. Quite simply, many people are just too busy to get around to much extra reading about possible new methods. Alternatively, many disciplines show extraordinary willingness to experiment with esoteric new techniques, particularly this season's hottest (or coolest) modeling approaches. At the same time, they show extraordinary inertia over more mundane details of data analysis, especially descriptive or presentational methods.

Historians tell us that the details of selection have often been curious and protracted. Broadly speaking, their stories imply that the optimists and pessimists are both largely correct. Every really good idea in statistical science experienced a glorious take-off when it became widely appreciated. But in retrospect, we can also identify failed precursors, similar or even identical, that did not really get off the ground. The bootstrap, which in a strong sense sprang from the head of Bradley Efron (1979), was anticipated by quite a few other ideas (Hall 2003). The box plot, which is often attributed to John W. Tukey

(1972, 1977), has a longer history stretching over at least a few decades (e.g., Crowe 1933).

Prosaically, even good simple ideas may need a distinct push to get them moving. The push sometimes requires a lot of energy: every field has its enthusiasts who tirelessly promote a method until people start to take notice. Kenneth E. Iverson (1920–2004) was the principal architect of APL and J and a major influence on many mathematical and statistical programming languages, including Stata. He was well advised by the Harvard computer pioneer Howard H. Aiken (1900–1973): "Don't worry about people stealing your ideas. If they're any good, you'll have to ram them down their throats" (Falkoff and Iverson 1981, 682).

The push is sometimes levered by a clever sales pitch, even down to the marketing details of a good name for the method. "Bootstrap" and "box plot" are cases in point, each hinting strongly at something simple and practical, but interestingly different from what you may know about already. Who can summon up much enthusiasm for learning about dispersion diagrams, the term under which climatologists and geographers used precursors of the box plot? The very name sounds dull and dreary.

Appositely for this journal, the push is sometimes eased by the existence of accessible software. Some discussions of software for data analysis seem based on the premise that trapped inside the body of every data analyst there is a programmer longing to get out. It is perhaps closer to the truth to say that most data analysts would much prefer that someone else did the programming, and for many good reasons.

This fairly simple analysis of a general issue provides context for one of my recurrent themes: methods that somehow or other lurk beyond the bounds of what is standard, perhaps for no good reason except the lack of a suitable push. In the next few columns I will look at some personal favorites, writing about them and their Stata implementations.

2   Empirical and theoretical densities

We start with density probability plots, a graphical method for examining how well an empirically derived density function fits a theoretical density function for a specified probability distribution. Here we are focusing on variables measured on scales that are at least approximately continuous. The terminology of density probability plots may suggest to you that, by accident or perversity, a nonstandard name is being used for what will turn out to be a standard kind of graph, but these plots are not standard at all. They appear to have been used very little in the decade since they were invented (Jones and Daly 1995). We will look first at existing methods for the overall problem. Then we will see better where density probability plots fit into the toolkit.

2.1   Histograms and kernel estimates

Suppose that we wish to compare the observed distribution of a variable Y with a specified distribution X with density function $f(x)$. To the density function, there correspond a distribution function $P = F(x)$ and a quantile function $X = Q(p)$. The distribution function and the quantile function are inverses of each other.

The particular density function $f(x|\text{parameters})$ most pertinent to comparison with data will be computed, given values for its parameters, either estimates from data or values chosen for some other good reason. For example, the parameters of a normal (Gaussian) distribution would usually be the mean and the standard deviation. We might estimate the mean and standard deviation, or we might just want to compare a distribution with a standard normal with mean 0 and standard deviation 1. As good courses and texts explain, such density functions are often superimposed on histograms or compared with density estimates produced using a kernel method. Such comparison amounts to comparing a density function fitted or estimated *globally* (say, a normal with particular mean and standard deviation) with a density function fitted or estimated *locally* (say, densities estimated for particular bins or by a moving kernel).

Spelling out the logic, which is simple and usually informal, there are three key ideas:

1. If the distribution model is a good fit, global and local density estimates will be close, but not otherwise.

2. Assessing what is acceptably close may pose a difficult judgment call, but assessing which of two or more candidate distributions is closer to a representation of sample data is often much easier. That is, it will often be easier to say whether a dataset is closer to a normal or to a gamma or a lognormal than it is to say whether a dataset is close enough to a normal to be declared as such.

3. The procedure is often useful for indicating *how* a model fails to fit, for example, in one or both tails or because of outliers. In particular, this is often clearer from a graph than as a by-product of any formalized or automated attempt to attack the problem as one of measuring degree of fit or one of testing a specific hypothesis.

As a matter of history, over most of the last century or so, histograms have been far more commonly used than kernel estimates. This is not really because kernel estimates are a more recent idea. Both ideas can be traced to the 19th century, kernel estimation to at least as far as C. S. Peirce in 1873 (Stigler 1978). The main inhibition has been that, until very recently, histograms were much easier to produce.

These procedures will be familiar to most readers, but let us remind ourselves how to do it in Stata. Manual references are [R] **histogram**, [G-2] **graph twoway histogram**, [R] **kdensity**, and [G-2] **graph twoway kdensity**. We will start with problems in which we think we know the answer, taking samples of random numbers from normal and exponential distributions. For reproducibility, set the seed before selecting samples:

```
. set seed 20
. set obs 500
. generate normal_sample = invnormal(uniform())
. label variable normal_sample "normal sample"
. generate exp_sample = -ln(uniform())
. label variable exp_sample "exponential sample"
```

Note that I used the Stata 9 function name `invnormal()` rather than the name used in previous versions of Stata, `invnorm()`. `invnorm()` continues to work in Stata 9. Readers still using earlier versions will need to use `invnorm()` to replicate this example. Focusing on the sample from the normal, we get the graphs combined in figure 1.

```
. histogram normal_sample, color(none) start(-4) w(0.2)
> plot(function normal = normalden(x,0,1), ra(-4 4))
. kdensity normal_sample, biweight w(0.5) n(500)
> plot(function normal = normalden(x,0,1), ra(-4 4))
```

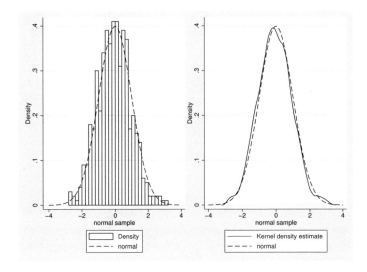

Figure 1: Histogram and kernel estimates of density of a sample of 500 values from a standard normal distribution, shown superimposed

Such an example highlights various familiar points. First, both histograms and kernel estimation require choices, respectively of bin origin and bin width, and of kernel type and kernel width. In particular, if we smooth too much, we throw away detail that might be informative, while if we smooth too little, we might be distracted by detail that is not informative. The choice is often awkward, even with much experience and a well-behaved dataset. Second, we are reminded that even with a moderate sample size—and in this case what should be an excellent fit—empirical distributions can show irregularities that may or may not be diagnostic of some underlying problem.

2.2   Quantile–quantile plots

A common answer to the first point, and to some extent even of the second point, is
to think in terms of not densities but quantiles and to produce a probability (meaning,
quantile–quantile) plot of observed quantiles versus expected quantiles. The Stata command qnorm does this for the Gaussian. See [R] **diagnostic plots**. With our sample,
this yields figure 2.

```
. qnorm normal_sample, ms(oh)
```

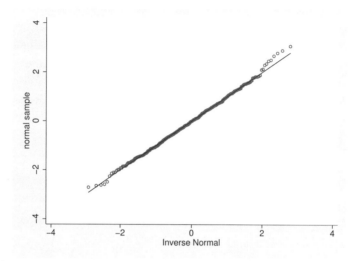

Figure 2: Quantile–normal plot of a sample of 500 values from a standard normal
distribution

In this case, other names used are normal quantile plot and probit plot. qnorm is
not *exactly* what we most need here, as it always uses the mean and standard deviation
of the data to fit a normal, rather than externally specified parameters (here mean 0
and standard deviation 1), but that is in practice just a very fine distinction.

Many experienced data analysts would regard the quantile–quantile plot as optimal
for this problem. There are almost no awkward choices to be made. The main, and
relatively minor, issue is which probability levels or plotting positions are used to evaluate expected quantiles; that is, what p to use in invnormal(p). In addition, plotting
observed and expected quantiles on a scatter plot reduces the problem of comparison
to one of assessing how far points cluster near the line of equality. This is easier than
comparing a curve with a curve. The value of linear reference patterns was emphasized
in a previous column (Cox 2004a).

The great merits of quantile–quantile plots are not in dispute here. They are excellent for assessing fit graphically, the task for which they were designed. However, they
leave the common desire for a visualization of densities unanswered. This then is the
setting for introducing density probability plots.

3   Density probability plots

The density function can also be computed via the quantile function as $f(Q(P))$. Parzen (1979) called this the density-quantile function. (See Parzen [2004] for an overview of his work on quantiles.) For example, if P were 0.5, then $f(Q(0.5))$ would be the density at the median. In practice, P is calculated as the plotting positions p_i attached to values $y_{(i)}$ of a sample of size n, which have rank i; that is, the $y_{(i)}$ are the order statistics $y_{(1)} \leq \cdots \leq y_{(n)}$.

Focusing now on that detail, one simple rule uses $p_i = (i - 0.5)/n$. Most other rules follow one of a family $(i - a)/(n - 2a + 1)$ indexed by a. Although there is literature agonizing about the choice of a, it makes little difference in practice. I tend to use $a = 0.5$ so that $p_i = (i - 0.5)/n$. This can be explained simply as splitting the difference between i/n, which can render the maximum unplottable, as it corresponds to plotting position 1, and $(i - 1)/n$, which can create a similar problem with the minimum, corresponding to plotting position 0.

For our normal example, denote by $\phi()$ and $\Phi()$, as usual, the theoretical density function and distribution function. In Stata 9, `normalden()` is the name for the function previously called `normden()`. The old name continues to work and should be employed by readers still using older versions.

We can thus compute

1. global estimates of the density if normal, given a specified mean and standard deviation, namely, $\phi\{(X - \text{mean})/\text{sd}\}$. In Stata terms, this is `normalden(`x`,mean,sd)`, and

2. local estimates of the density, namely, $\phi\{\Phi^{-1}(p_i)\}$. In Stata terms, this is `normalden(invnormal(`p`))`.

As with a quantile–quantile plot, there is only one key graphical decision, the choice of plotting positions. As with previously existing methods, the logic is that if the distribution model is a good fit, global and local density estimates will be close, but not otherwise. The match between the curves allows graphical assessment of goodness of fit and identifying location and scale differences, skewness, tail weight, tied values, gaps, outliers, and so forth.

Such density probability plots were suggested by Jones and Daly (1995). See also Jones (2004). They are best seen as special-purpose plots, like normal quantile plots and their kin, rather than general-purpose plots, like histograms or dot plots. In a program, each distribution fitted (gamma, lognormal, whatever) needs specific code for its quantile and density functions. A Stata implementation is given in the `dpplot` program, published with this column. Fitting a normal is the default. Other distributions that may be fitted include the beta, exponential, gamma, Gumbel, lognormal, and Weibull: some of these require programs to be downloaded from SSC or else a program to be written by the user if parameters are to be estimated from the data, as will usually be desired. In addition, Stata users with some programming expertise should be able to modify existing programs to fit any favorite distribution that is not supported and also to write extra programs compatible with `dpplot` to carry out the necessary calculations.

For our normal problem, the minimum necessary is to specify the parameter values of interest:

```
. dpplot normal_sample, param(0 1) ms(oh)
```

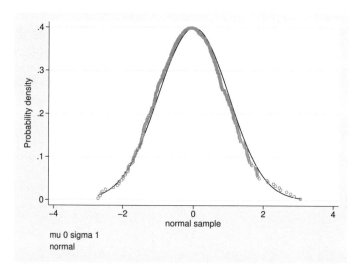

Figure 3: Density probability plot of a sample of 500 values from a standard normal distribution showing fit to that standard normal

Figure 3 gives another graphical take on the problem. The fit would usually be declared good, but it takes some experience, or some theory, to know that the slight irregularities in the tails are not enough to worry about.

Turning to the exponential sample, the essential logic is identical. We need to plug in a value for the mean, the single parameter, to compute

1. global estimates of the density if exponential, namely, $(1/\text{mean}) \exp(-X/\text{mean})$. In Stata terms, this is `(1/mean) * exp(-X/mean)`, and

2. local estimates of the density. The quantile function is

$$X = -\text{mean} \ln(1 - P)$$

and the density function, as above, is

$$(1/\text{mean}) \exp(-X/\text{mean})$$

Telescoping the two, this is

$$(1/\text{mean}) \exp\left[-\{-\text{mean}\ \ln(1-P)\}/\text{mean}\right]$$

which reduces easily to $(1-P)/\text{mean}$.

Our mean is tacitly 1, so this is all done for us by one command, yielding figure 4.

```
. dpplot exp_sample, dist(exp) param(1) ms(oh)
```

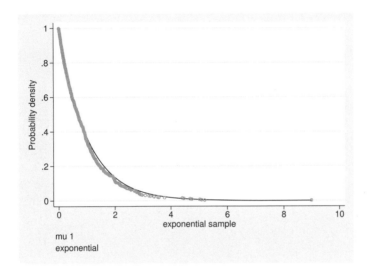

Figure 4: Density probability plot of a sample of 500 values from an exponential distribution with mean 1 showing fit to that distribution.

In this case, the local estimates droop somewhat below the theoretical curve. The fact that the sample mean is rather lower than 1 at 0.942 is pertinent here. Every now and then, our random sample will not match the ideal well, "just by chance".

4   An example with real data

Let us look at an example with real data, on 158 glacial cirques from the English Lake District (Evans and Cox 1995). These were examined in a previous column (Cox 2004b). Glacial cirques are hollows excavated by glaciers that are open downstream, bounded upstream by the crest of a steep slope (wall), and arcuate in plan around a more gently sloping floor. More informally, they are sometimes described as "armchair-shaped". Glacial cirques are common in mountain areas that have or have had glaciers present. Three among many possible measurements of their size are length, width, and wall height. As a first stab at fitting specific distributions, we consider three two-parameter distributions, namely,

1. The gamma distribution is parameterized using a scale parameter β and a shape parameter α so that the density function is

$$\frac{1}{\beta^{\alpha}\Gamma(\alpha)} x^{\alpha-1} \exp\left(\frac{-x}{\beta}\right)$$

Here $\Gamma(u)$ is the gamma function $\int_0^{\infty} t^{u-1} \exp(-t)\,dt$.

2. The lognormal distribution is parameterized using a location parameter, the mean of the logs μ, and a scale parameter, the standard deviation of the logs σ, so that the density function is

$$\frac{1}{X\sigma\sqrt{2\pi}} \exp\left\{ -(\ln X - \mu)^2/2\sigma^2 \right\}$$

3. The Weibull distribution is parameterized using a scale parameter β and a shape parameter γ so that the density function is

$$\frac{\gamma}{\beta}\frac{x^{\gamma-1}}{\beta} \exp\left\{ -(x/\beta)^{\gamma} \right\}$$

Note that these parameterizations all differ from those employed by `streg`. See [ST] **streg**. In each case, the distribution was fitted using maximum likelihood.

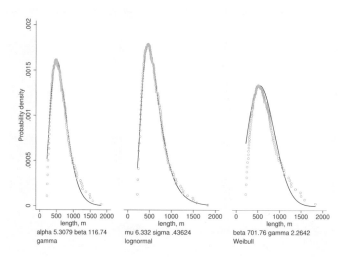

Figure 5: Density probability plots showing fits of cirque length data to gamma, lognormal, and Weibull distributions

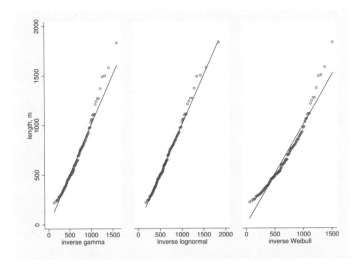

Figure 6: Quantile–quantile plots showing fits of cirque length data to gamma, lognormal, and Weibull distributions

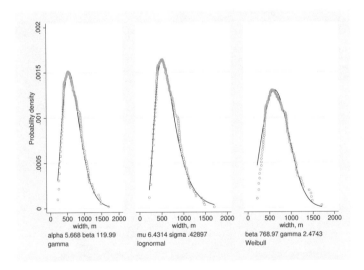

Figure 7: Density probability plots showing fits of cirque width data to gamma, lognormal, and Weibull distributions

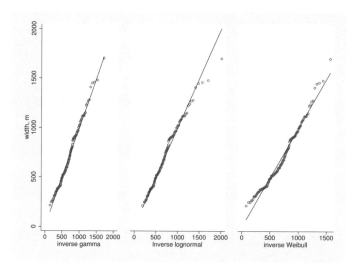

Figure 8: Quantile–quantile plots showing fits of cirque width data to gamma, lognormal, and Weibull distributions

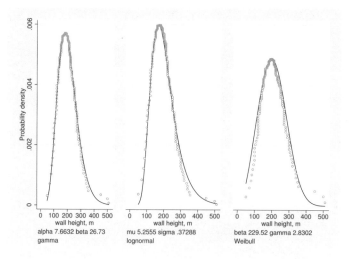

Figure 9: Density probability plots showing fits of cirque wall height data to gamma, lognormal, and Weibull distributions

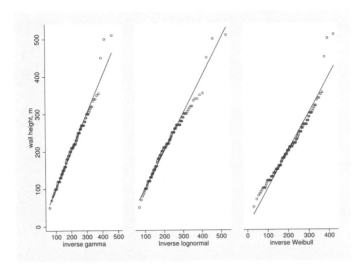

Figure 10: Quantile–quantile plots showing fits of cirque wall height data to gamma, lognormal, and Weibull distributions

Figures 5 through 10 show the results of fitting these distributions to the three size variables, using both density probability plots and quantile–quantile plots. The Weibull is a poor fit in every case, so we will say no more about it. In the fight between the gamma and the lognormal, the latter seems to have the edge for length (figures 5 and 6) and the former for width (figures 7 and 8) and wall height (figures 9 and 10). Three moderate outliers for wall height are enigmatic, however they are viewed.

This is not a complete story. Many would feel it worth exploring three-parameter distributions bringing in nonzero thresholds. Cirques cannot be arbitrarily small (i.e., arbitrarily close to zero in size), as the glaciers that form them cannot be: tiny snow patches do not become glaciers without growing first. But the main message here is one of method, suggesting that density probability plots can nicely complement quantile–quantile plots.

5   The modality of the distribution fitted is always shown

Many users of density probability plots are surprised by one of their features. A little thought shows that this feature is an inevitable consequence of what is being requested, but it is worth explanation.

Suppose that we do something rather silly: we plot the exponential sample to see how far it is normal. To mix silliness and sense, we will use the theoretical principle that the mean and standard deviation should both be 1:

```
. dpplot exp_sample, param(1 1) ms(oh)
```

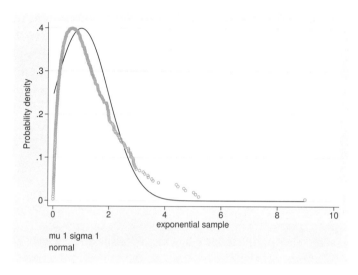

Figure 11: Density probability plot of a sample of 500 values from an exponential distribution with mean 1 showing fit to a normal distribution with mean 1 and standard deviation 1

The plot in figure 11 shows that the fit is poor, but that should be a foregone conclusion. The surprise may be that the empirical or local density estimate is strongly unimodal with definite tails on both left and right. It does not look Gaussian, but even less does it look exponential. How does that shape arise?

The key here is that what you are seeing is the exponential sample trying its very hardest to look as normal (Gaussian!) as possible. That is, you are seeing a normal density, but continuously transformed. When data points are closer together, or further apart, than they would be if the data were really normal, then the normal is also squeezed or stretched accordingly. But those deformations affect horizontal position only, and the density ordinates are exactly as calculated by plugging into `normalden()`. The modality remains that of the normal distribution being fitted.

More generally with density probability plots, results may be difficult to decipher if observed and reference distributions differ in modality or other aspects of gross shape. So if the density function of the distribution being fitted is monotone decreasing or unimodal with two tails, then so too must be $f(Q(P))$, whatever the shape of $f(Y)$.

The practical implication is that when the fit between empirical and theoretical densities is really poor, a density probability plot may well be obscure on quite why and how the fit is poor. The last example was silly on purpose, but there is little to stop even experienced data analysts doing silly things by accident. In particular, density probability plots are not especially suitable for very naive users.

6   Virtues and vices

Extending the discussion in Jones and Daly (1995), the advantages of these plots include

- *Ease of interpretation.* Some people find them easier to interpret than quantile–quantile plots.

- *Fewer awkward choices.* No choices of binning or origin (cf. histograms, dot plots, etc.) or of kernel or of degree of smoothing (cf. density estimation) are required.

- *Flexibility with regard to sample size.* They work well for a wide range of sample sizes. Nevertheless, just as with all other methods, a sample of at least moderate size is preferable (say, ≥ 25).

- *Bounded support is clear.* If X has bounded support in one or both directions, then this should be clear.

The disadvantages include

- *Modality may not match.* Results may be difficult to decipher if observed and reference distributions differ in modality.

- *Tails may be cryptic.* It may be difficult to discern subtle differences in one or both tails of the observed and reference distributions. On the other hand, tails are not always crucial, and it is arguable that quantile–quantile plots may have the opposite weakness of overemphasizing tails.

- *Comparison of curves.* Comparison is of a curve with a curve: some people argue that graphical references should, where possible, be linear (and ideally horizontal). On the other hand, `dpplot` has a `generate()` option to save its results and a `diff` option to show the difference between the two guesses at the density function.

- *Not extensible.* There is no simple extension to comparison of two samples with each other.

7   Conclusions

Density probability plots use sample data to show how close a data sample is to a specified distribution. They portray two guesses at the density function, one produced globally, usually by estimating the parameters of the distribution, and one produced locally by evaluating the density-quantile function at the data points. Although proposed a decade ago, they appear to have been used very little, and this column has provided further publicity, a pointer to a Stata implementation, and a discussion of their advantages and limitations. In particular, they avoid the choices of bin or kernel width imposed by histograms or kernel density estimation, and they usefully complement quantile–quantile plots in conveying what distribution is being fitted, as well as how well that distribution fits.

8   Acknowledgments

Ian S. Evans, Stephen Jenkins, and Tim Sofer aided this project in various ways.

9   References

Cox, N. J. 2004a. Speaking Stata: Graphing agreement and disagreement. *Stata Journal* 4: 329–349.

————. 2004b. Speaking Stata: Graphing distributions. *Stata Journal* 4: 66–88.

Crowe, P. R. 1933. The analysis of rainfall probability: A graphical method and its application to European data. *Scottish Geographical Magazine* 49: 73–91.

Efron, B. 1979. Bootstrap methods: Another look at the jackknife. *Annals of Statistics* 7: 1–26.

Evans, I. S., and N. J. Cox. 1995. The form of glacial cirques in the English Lake District, Cumbria. *Zeitschrift für Geomorphologie* 39: 175–202.

Falkoff, A. D., and K. E. Iverson. 1981. The evolution of APL. In *History of Programming Languages*, ed. R. L. Wexelblat, 661–691. New York: Academic Press.

Hall, P. 2003. A short prehistory of the bootstrap. *Statistical Science* 18: 158–167.

Jones, M. C. 2004. Hazelton, M. L. 2003, "A graphical tool for assessing normality", American Statistician 57: 285–288: Comment. *American Statistician* 58: 176–177.

Jones, M. C., and F. Daly. 1995. Density probability plots. *Communications in Statistics—Simulation and Computation* 24: 911–927.

Parzen, E. 1979. Nonparametric statistical data modeling. *Journal of the American Statistical Association* 74: 105–131.

————. 2004. Quantile probability and statistical data modeling. *Statistical Science* 19: 652–662.

Stigler, S. M. 1978. Mathematical statistics in the early states. *Annals of Statistics* 6: 239–265.

Tukey, J. W. 1972. Some graphic and semigraphic displays. In *Statistical Papers in Honor of George W. Snedecor*, ed. T. A. Bancroft and S. A. Brown, 293–316. Ames, IA: Iowa State University Press.

————. 1977. *Exploratory Data Analysis*. Reading, MA: Addison–Wesley.

About the author

Nicholas Cox is a statistically minded geographer at Durham University. He contributes talks, postings, FAQs, and programs to the Stata user community. He has also co-authored fifteen commands in official Stata. He was an author of several inserts in the *Stata Technical Bulletin* and is an Editor of the *Stata Journal*.

The Stata Journal (2005)
5, Number 3, pp. 442–460

Speaking Stata: The protean quantile plot

Nicholas J. Cox
Durham University, UK
n.j.cox@durham.ac.uk

Abstract. Quantile plots showing by default ordered values versus cumulative probabilities are both well known and also often neglected, considering their major advantages. Their flexibility and power is emphasized by using the `qplot` program to show several variants on the standard form, making full use of options for reverse, ranked, and transformed scales and for superimposing and juxtaposing quantile traces. Examples are drawn from the analysis of species abundance data in ecology. A revised version of `qplot` is formally released with this column. Distribution plots in which the axes are interchanged are also discussed briefly, in conjunction with a revised version of `distplot`, also released now.

Keywords: gr0018, qplot, distplot, distributions, quantile plots, statistical graphics, species abundance, ecology, Whittaker plots, broken stick, lognormal, power laws, scaling laws

1 Introduction

In my last column (Cox 2005), I started to look at some personal favorite methods that somehow or other lurk beyond the bounds of what is standard. The first method examined was that of density probability plots, which have attracted very few users in the decade since they were proposed. For this column, we stay in the same territory of distribution graphics. The method to be examined now is that of quantile plots for showing one or more distributions. Regarding these as not standard may appear more disputable. The advantages of quantile plots are first discussed, making clear that they provide solutions to the main limitations of the much more popular histograms and box plots. Examples of standard ecological analyses are then used to show how versatile quantile plots can be.

Quantile plots graph a set of ordered values[1] against the so-called plotting positions, in essence the associated cumulative probabilities. They may well be in the routinely used toolkit of several readers. Quantile plots have been described in detail in various leading texts (e.g., Chambers et al. 1983; Cleveland 1993, 1994). An official Stata implementation has been available for some time in `quantile`. An extended user-written implementation has also been available in `qplot`, formerly `quantil2` (Cox 1999b). Moreover, I have previously discussed quantile plots together with other methods for graphing distributions (Cox 2004c). What then are the grounds for regarding quantile plots as neglected?

---

1. Otherwise put, the (sample) quantiles for our purposes are identical to the order statistics. This terminology may seem loose if you are accustomed to other definitions of quantiles, but it is standard in the statistical graphics literature. For a review of order statistics, see David and Nagaraja (2003).

2   Quantile plots are needed

In broad terms, the case is that quantile plots are even more useful than is widely realized. Specifically, this column coincides with the release of a further-enhanced version of `qplot`. More generally, there are several grounds for promoting their use. The arguments differ to some extent, depending on whether teaching or research is the main issue.

At an introductory level, the most-popular methods taught for showing univariate distributions of variables that are continuous, or nearly so, appear to be histograms and box plots. Dot plots are a popular third alternative in some quarters. The choice of what to teach is usually guided by a need to explain methods that may be encountered by students or practitioners in their later courses or subsequent careers, so selection is often conservative.

The merits of histograms and box plots do not need much emphasis, but their limitations do require brief mention. Without saying everything that could be said, here are some standard comments.

1. The binning that is the basis of histograms is often awkward. Every histogram depends on decisions about bin width and origin. Texts thus have to explain the compromise required between undersmoothing that omits too little detail and oversmoothing that omits too much. Although learners do not find the principle difficult to understand, in practice they often make poor choices with their own histograms. Arguably, good choice of bin width depends on statistical experience—precisely what introductory audiences rarely have. Without that, there is much scope for small or large misinterpretations over key detail in the tails, real and imagined modality, and even gross distribution shape. For example, slight but notable skewness can easily be missed with poor bin choices.

2. Histograms can be excellent for individual distributions, but too often they tend to be dead ends leading nowhere else. Overlay of a curve (especially some theoretical or fitted density function) can be very useful, but the comparison of two or more histograms is frequently difficult. Neither juxtaposition (histograms side by side) nor superimposition (histograms on top of each other) usually works well, except for the crudest propaganda purposes and a small number of histograms. (Look! The distributions are almost the same! Or: Look! The distributions are very different!) Ask yourself: How many histograms could you superimpose and still have your readers come to quantitative conclusions easily and effectively? How many histograms could you juxtapose with the same result?

3. Conversely, box plots can be excellent for comparisons but arguably have been oversold for showing single batches of the data. With, say, dozens of groups in the data, the condensation imparted by box plots can be vital for seeing overall patterns together with gross anomalies. With one or just a few groups of data, box plots can show too little about what is going on both in the central box and in the tails. This is one reason for the steadily growing popularity of dot plots.

The limitations of these most popular choices correspond to the strengths of quantile plots.

1. Quantile plots require no decisions about bin width or origin or similar matters. There is scope for varying axis scales from the default, but this is a feature, not a problem to be solved.

2. Quantile plots show empirical curves that are weakly monotonic. These are relatively easy to compare, certainly considering the difficulties of comparing histograms. Typically, users choose point or line representations. In Stata, the flexibility possible over plottype arises through the scope for using `recast()` with `quantile` or `qplot` to produce a different kind of graph. Moreover, in the latest version of `qplot`, the user can take advantage of either or both an `over()` option and a `by()` option to show superimposed and juxtaposed traces as desired.

3. Quantile plots in principle show all the information in the distribution but without either emphasizing or suppressing any particular data points. Thus fine structure in the tails or indeed anywhere else is evident. Contrast the case of histograms, which may suppress much or all of the fine structure through their use of bins, and the case of box plots, which may flag apparent outliers but suppress virtually all other detail.

Unsurprisingly, quantile plots have disadvantages, too. We make two comments now and will say more later.

1. For some groups, they may seem unfamiliar, but this is precisely where we came in. You need to learn how to "read" quantile plots, just as you do any other kind of graph. Managers, advisors, or editors (but not the Editors of the *Stata Journal*, naturally) might insist that you use what is to them a more familiar kind of graph, but exercise of power is not the best way to win an argument.

2. More substantially, if your problem is better illuminated by looking at densities, as when the modality of a distribution is of concern, a density trace (such as that produced by `kdensity`) will clearly be more appropriate.

The definition of quantile plot here has been quite strict: we insist that quantiles are plotted as response on the vertical axis. Many readers would be happy with the idea that distribution plots, in which cumulative probability or frequency is plotted on the vertical axis versus quantiles on the horizontal axis, are just a variant on quantile plots, with the main difference that the axes are interchanged. More will be said on this point later.

More broadly, emphasis on quantiles rather than cumulative distributions is more than a matter of cosmetics or aesthetics. The tendency persists in many quarters of introducing the quantile function rather grudgingly and indirectly as the inverse of the distribution function. Authors are perhaps here recapitulating the historical order, rather than following the most natural scientific route. Be that as it may, whenever the quantiles are essentially the response of interest, they belong on the vertical axis.

3   Abundance distributions in ecology

A key problem in ecology is to compare counts from several sites of the numbers of individuals in various different species. Almost always, the attention is not to all organisms present but is restricted to groups of fairly closely related species, say, birds, lizards, fish, or trees. Typically, there is much variation from a few very common species to rather more (even very many more) rarer species at each site. How can we best report such data in terms of distributions fitted or informative summary measures? This is usually preliminary to some kind of assessment of relationships with other ecological or environmental variables.

This can be seen as the prototype for several related problems. Focus at the species level is common, but not universal, as other taxa (units on other taxonomic levels, say, genera or subspecies) could be used. Counting individuals is often the most straightforward method, but with some organisms (e.g., corals, mosses), the definition of an individual can be problematic, and in any case, it may be desired to measure surface cover, biomass, productivity, or some other property. In our examples, we will focus on counting. Little imagination is needed to appreciate that even at restricted sites and with appropriate taxonomic skills, ecologists generally face a sampling problem, as some species may never be seen within a particular sample. Only in a few cases, say, recording all trees present at a temperate forest site, is anything like a complete census possible.

This basic problem is part of the descriptive science underlying the analysis of ecological diversity, or biodiversity, that is so prominent in many environmental discussions. The literature is enormous. Magurran (2004) offers a fairly informal introduction with worked examples. Lande, Engen, and Sæther (2003) relate the analysis of diversity to population dynamics at a deeper level. ? remains an outstanding blend of ecological, mathematical, and statistical thinking. Hutchinson (1978) tastefully embeds the story in the wider history of ideas.

An ecological problem is being used here as a vehicle for some graphical ideas, but no discipline possesses totally idiosyncratic questions. In particular, there are many parallels with problems in what ostensibly is a very different field, analysis in economics of the distribution of incomes, firm sizes, and so forth. Kleiber and Kotz (2003) give an excellent review of the literature. In each case, much of the challenge lies in the handling of very skewed and long-tailed distributions. In each case, problems of sampling, censoring, and truncation can be acute. The jargon in ecology is of diversity, species richness, and species evenness, and in economics of inequality and concentration, but the issues remain sisters under the skin. The pitfalls too show resemblances. Much time and effort have gone in each discipline into searches for summary measures (indexes, coefficients, and what have you) that somehow capture the interesting information, remain interpretable theoretically and are robust to whatever investigators regard as perversities in their data. In each case, skeptics intermittently inject notes of caution about such an enterprise, while others keep on hoping. Despite these similarities, cross-fertilization of literatures has been very slight[2]. What is more common is multiple rediscovery of the same methods, or ignorance of relevant solutions elsewhere. What to ecologists is the Simpson index is to economists the Herfindahl index and to yet others one of the many beasts called a Gini index, to give one small example.

4 Birds in the Killarney woods

Let us consider the best structure in Stata for ecological abundance data. If only presence is recorded, each observation might as well be site name, species name, and number of individuals. Thus we represent frequencies in a long structure, to use the terminology of [D] **reshape**. A wide data structure with either each site as a variable, or each species as a variable, has some advantages for another analyses but can be decidedly awkward for graphics, as we are likely to hit against limits on the number of variables that can be shown on a graph. In some problems, the species names can be ignored—or were never supplied in literature reports—so that data take on the form of frequencies of frequencies, that is, how many species are represented by 1, 2, 3, … individuals.

4.1 Whittaker plots

A particular variant on the quantile plot is often now recommended by ecologists as the best single starting point for showing data of this kind. Abundance on a logarithmic scale should be plotted against the rank of each species. Krebs (1989, 344) called such a plot a *Whittaker plot* after the leading American ecologist Robert Harding Whittaker (1920–1980), who used them prominently. Ecologists follow the opposite convention from statisticians and assign rank 1 to the most common species. The fact that the least-common species may be difficult or even impossible to determine is grounds enough for this convention.

Figure 1 shows a simple example of a Whittaker plot. The data on bird species diversity come from Magurran (2004, 237–238), who in turn got them from a study by Batten (1976, 307, 309, 313) in contrasting woodlands in Killarney, Ireland. The rationale for the log scale is clear: even in this rather small and simple dataset, there is a 35-fold variation between the most- and least-common species. On a log scale, with rank on a linear scale, a geometric series would show as a straight line, and a logarithmic series distribution (Fisher, Corbet, and Williams 1943) as very nearly a straight line. On the evidence here, there is some support for going further with either of those models, but also a worry that the dataset is rather small.

---

2. Simon (1991, 367) wrote that "Disciplines, like nations, are a necessary evil that enable human beings of bounded rationality to simplify their goals and reduce their choices to calculable limits".

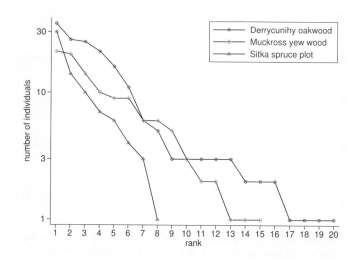

Figure 1: Whittaker plot showing bird diversity in three woodlands in Killarney, Ireland. The vertical log scale gives a good first look at the distributions.

The command for this plot is

```
. qplot count, over(site) rank reverse recast(connected) yscale(log)
> ytitle(number of individuals) ylabel(1 3 10 30, angle(h)) xlabel(1/20)
> legend(position(1) ring(0) column(1))
```

The `over(site)` option gives multiple traces, one for each site, in each panel shown; here just one is shown. An alternative, not so helpful here, would be `by(site)`, giving curves for each site in separate panels. The `rank` and `reverse` options spell out the scale for the horizontal axis. The other options are just standard `twoway` options.

A common alternative is to show abundance as a percent of each site total (figure 2). That requires just a prior `egen` call:

```
. egen pc = pc(count), by(site)

. qplot pc, over(site) rank reverse recast(connected) ysc(log)
> yti(percent of total) yla(1 3 10 30 100, ang(h)) xla(1/20) legend(pos(1)
> ring(0) col(1))
```

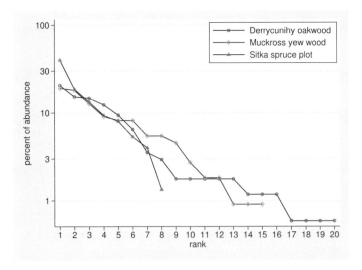

Figure 2: Whittaker plot showing bird diversity for three woodlands in Killarney, Ireland. The vertical scale shows percent of total abundance.

Clearly, proportions between 0 and 1 are possible, as well as percentages.

4.2   Broken stick distributions

Several distributions have been entertained in the ecological literature for diversity analysis, in addition to the logarithmic series mentioned in passing above. Extra interest attaches to those with some more or less plausible ecological rationale or interpretation, at least as a kind of null or reference model. We will look in the first instance at two such distributions, the so-called broken stick and the lognormal, to show how quantile plots (and `qplot` in particular) can be used in initial exploration of the applicability of some models. One key point that will be understated is the extent to which the distributions mentioned are special or limiting cases of other more flexible distributions, such as the gamma or the negative binomial (e.g., Lande, Engen, and Sæther 2003).

The name of the broken stick arises from one way of visualizing the splitting of a whole randomly into parts, each part here corresponding to one of S species. Suppose that a stick of unit length is broken at random in $S - 1$ places, thus yielding S parts, whose lengths we order as $p_1 \geq p_2 \geq \cdots \geq p_S$. The mean lengths of these ordered proportions p_i can be shown to be

$$p_i \;=\; \frac{1}{S} \sum_{r=i}^{S} \frac{1}{r}$$

and so, given a total of N individuals, the expectation is of

$$Np_i = \frac{N}{S} \sum_{r=i}^{S} \frac{1}{r}$$

individuals in the ith most common species. Note that there are no parameters to be estimated, an interesting detail. This fact naturally does not remove all statistical flavor from the problem: the breaking of the stick is probabilistic, so there will be variance around each mean, a point deserving more attention in a fuller account.

This general rule can be cross-checked against other arguments for the simplest interesting case, splitting into $S = 2$ species. The two parts have average proportions

$$p_1 = (1 + 1/2)/2 = 0.75 \qquad \text{and} \qquad p_2 = (1/2)/2 = 0.25$$

which matches an arm-waving argument that on average the break will be midway from the end of the half in which it occurs. Mosteller (1965, 62–65) goes carefully through the cases of $S = 2$ and $S = 3$ and gives the general result. Cohen (1966) is an elegant monograph (a reworking of the author's Bachelor's dissertation) linking the broken stick to ecological and economic theory and examples.

The broken-stick distribution has been rediscovered in various ways over the last century or so, and you may know it under other names. The trail goes back at least as far as William Allen Whitworth (1840–1905) and his textbook *Choice and Chance* (Whitworth 1905). Whitworth studied mathematics at Cambridge and then taught the subject in schools and University College, Liverpool. His later career developed as a minister in the Church of England. Whitworth's other books range from *Trilinear Coordinates* to the posthumous collection of sermons *The Sanctions of God*. For other information, see Irwin (1967) and Domb (1990).

The calculation of the broken-stick frequencies can be achieved in a few lines of Stata, but some small details deserve commentary.

```
. gsort site -count
. by site: generate double bstick = 1/_n
. by site: egen N = total(count)
. by site (bstick), sort: replace bstick = N * sum(bstick)/_N
```

The reciprocals of ranks $1/r$ are to Stata `1/_n`, so long as we have the right sort order. Recall that ecologists use the reverse-ranking convention, so that `gsort` is used to get the sort order needed, the minus sign specifying sorting from largest downwards. `double` is a precaution, as the reciprocals can be rather small and we want to carry as much precision as possible.

There is a clash between Stata's notation and that customary in the ecological literature. As each observation records a species frequency, the number of species S is given directly by `_N`. The number of individuals N is obtained by an `egen` call. From Stata 9, what used to be `egen, sum()` has been renamed `egen, total()` in an attempt to distinguish it more clearly from the `sum()` function.

The remainder of the calculation is a cumulative sum using `sum()`, together with division by S and multiplication by N. A delicate point here is that cumulation from the lowest ranking species upwards requires reversing the sort order from that produced by `gsort`. It is tempting to think (as I did in one stab at this problem) that `sort site count` reverses `gsort site -count` within each site, but that would overlook the effect of ties, which can be very common with this kind of data. Although tied frequencies will by definition be identical, the associated reciprocals `1/_n` will not be. Reversing by `sort site bstick` is safe.

If this were a routine problem, then naturally, thought would be given to encapsulating code in a program yielding broken-stick predictions directly.

A partial check on calculations is given by seeing whether observed and predicted frequencies have the same total:

```
. tabstat bstick count, s(sum) by(site)
```

This example of broken-stick calculation points up a standard Stata moral. You can do a lot of work with careful use of `by:`, `_n`, and `_N`, but you have to get exactly the right sort order for the problem. An earlier column (Cox 2002) discussed these features in more detail.

That done, we can now put the broken-stick quantiles on the plot. We have two variables, observed and fitted, and three sites, so we need to use a `by()` option to specify separate panels. We revert to the original scale for the quantiles (figure 3).

```
. qplot count bstick, by(site, row(1)) rank reverse recast(connected)
> yti(number of individuals) yla(, ang(h)) xla(1 5(5)20)
> legend(order(1 "observed" 2 "broken stick") pos(6) row(1) ring(0))
```

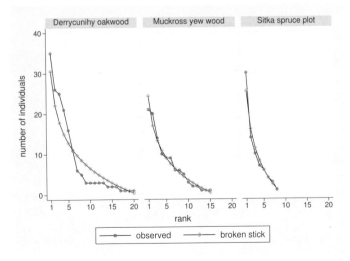

Figure 3: Broken sticks fitted for three woodlands in Killarney, Ireland. We revert to arithmetic scale for responses.

A good tip for broken sticks is that they are close to linear if plotted with a logarithmic scale for ranks (figure 4). The agreement I would describe as fair for the oakwood and better for the other two sites.

```
. qplot count bstick, by(site, row(1) note("")) rank reverse recast(connected)
> yti(number of individuals) yla(, ang(h)) xsc(log) xla(1/5 7 10 15 20)
> legend(order(1 "observed" 2 "broken stick") pos(6) row(1) ring(0))
```

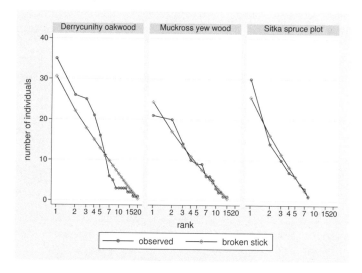

Figure 4: Broken sticks fitted for three woodlands in Killarney, Ireland. The horizontal log scale makes the broken sticks more nearly linear.

4.3   Lognormal distributions

Yet another commonly used distribution as a null or reference case is the lognormal. When fitting this distribution, ecologists most frequently draw histograms, often agonizing at some length about the choice of base of logarithms and how to bin. Whatever is chosen, there will be some loss of detail. Unsurprisingly, we remain with quantile plots and choose a log scale for the responses and an inverse-normal scale for the horizontal axis. This is done by specifying the `trscale()` option, which allows on-the-fly transformation of the horizontal-axis scale. @ is a place-holder for what would be plotted otherwise, probability or rank. Stata 8 users should here type `invnorm()`, not `invnormal()`. The online help for `qplot` explains how to get a labeling of the horizontal axis in terms of percent points, such as 1 2 5 10(10)90 95 98 99. Hydrologists would call such a display for stream discharges a flow-duration curve (e.g., Gordon et al. 2004).

```
. qplot count, over(site) reverse trscale(invnormal(@))
> xti(standard normal scale) recast(connected) ysc(log)
> yti(number of individuals) yla(1 3 10 30, ang(h))
> legend(pos(1) ring(0) col(1))
```

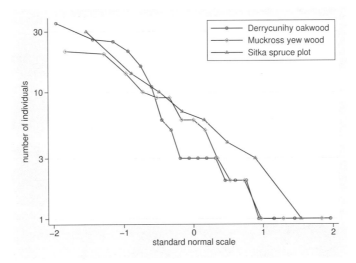

Figure 5: Lognormal distributions fitted for three woodlands in Killarney, Ireland. Note the vertical log scale and the horizontal inverse normal scale.

Again the fits look fairly good, leaving us with an embarrassment of choices and a feeling that we would need more data to decide confidently between these distributions. A further worry is whether we are being too cavalier in ignoring the discreteness of the counts. In particular, some ecologists have fitted Poisson lognormals (Bulmer 1974).

5   Macrolepidoptera at Rothamsted 1935 and the temptations of power laws

The main purpose here is to show off quantile plots, so leaving the previous example in midair does not feel too treacherous. Let us look at a more extensive dataset on macrolepidoptera (butterflies and the larger moths) caught in a light trap at Rothamsted (Hertfordshire, England) in 1935 and reported by Williams (1964, 25) (and also by Magurran 2004, 216). 197 species and 6815 individuals were caught, so the data are reported compactly as frequencies of frequencies; that is, 37 species are represented by a single individual, 22 species by two individuals, and so forth. We read them into Stata in that form and then used `expand count` to get an observation for each species.

The name Rothamsted will suggest to many readers the names of statisticians who have worked there at one time or another, including Sir Ronald Fisher, Frank Yates, and John Nelder. The overlap at Rothamsted of Fisher and the statistically minded entomologist Carrington Bonsor Williams (1889–1981) undoubtedly aided their collaboration[3]. This culminated in the key logarithmic series distribution paper of Fisher, Corbet and Williams (1943), one of the main roots of statistical ecology[4].

We will show just one histogram to make a point; it does show the main feature of this distribution clearly (figure 6). Moment-lovers may note calculated skewness of 10.51 and kurtosis of 127.99. The display is spectacular but useless. Not surprisingly, many ecologists would prefer a histogram based on log counts. We revert to quantile plots.

```
. histogram count, w(1) discrete freq yla(, ang(h)) yti(number of species)
> xti(number of individuals)
```

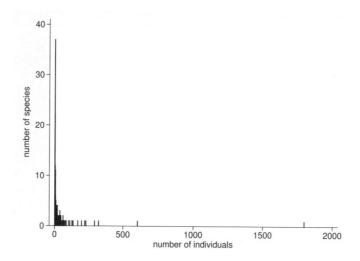

Figure 6: Macrolepidoptera at Rothamsted 1935. A highly skewed distribution for species with varying numbers of individuals.

With such an extreme distribution shape, we might even consider the idea of a power or scaling law. Over the last century or so, many scholars have fallen for the temptations of power laws. Economists may think of Vilfredo Pareto; linguists, sociologists, psychologists, and geographers of George Kingsley Zipf; physicists of Per Bak; and everyone numerate of Benoît Mandelbrot. The literature extends to grandiose works showing how

---

3. Letters from Fisher to Williams copied on the Internet at
   http://www.library.adelaide.edu.au/digitised/fisher/corres/williamscb/ show his nuanced progression over the years through greater degrees of informality: from "Dear Dr Williams", to "Dear Williams", "My dear Williams", and "Dear C.B." to "My dear C.B.". Notice the mutation through a variety of forms, none of which included the first names so thoughtfully bestowed by Williams' parents.

4. The third member of the trio, the British scientist Alexander Steven Corbet (1896–1948), started as a chemist and biochemist and ended as an entomologist, publishing the definitive work on the butterflies of the Malay peninsula. In between, he established the production of N_2O by denitrifying soil bacteria and wrote a book on biological processes in tropical soils (Riley 1948; Hutchinson 1978, 231).

power laws offer keys to understanding the universe, or at least some large segment of it, such as *Human behavior and the principle of least effort* (Zipf 1949) and *How nature works* (Bak 1997)[5]. For a level-headed discussion of power laws and how far they have been oversold, see Perline (2005).

In the simplest case, power laws have been suggested of the form

$$\text{count} \times \text{rank} = \text{constant}$$

or

$$\text{count} \propto \text{rank}^{-1}$$

although the greater generality of

$$\text{count} \propto \text{rank}^{-b}$$

has seemed an easy step further[6]. Here "count" is the right word, but other applications extend to measured sizes, such as earthquake magnitudes or avalanche volumes. Note again that this formulation depends on ranking the largest as 1.

The connection to quantile plots is immediate. With a power law of this form, quantiles expressed as a function of rank will plot as a straight line with downward slope b if both axis scales are logarithmic. For a variety of reasons, some mathematical and some empirical, a variety of further adjustments have been made to the form of the power law, but approximate linearity on a log-log plot remains the general expectation.

Figure 7 shows the result for our dataset. The enthusiast will note a substantial straightening; the skeptic will note remaining systematic curvature. (A regression of log frequency on log rank gives a downward slope of 1.59.)

```
. qplot count, reverse rank ysc(log) yti(number of individuals)
> yla(1 3 10 30 100 300 1000, ang(h)) xsc(log) xla(1 3 10 30 100)
```

---

5. Names must be part of the explanation. Suppose that I write a treatise on the negative binomial, using a wide variety of examples to show how it is astonishingly applicable in all sorts of problems across many sciences. Then I leave out most of the equations and add seasoning and garnish in the form of entertaining and poignant anecdotes and quirky connections. The project would never go public, as no publishing person would buy the notion that anything with such a bizarre name as "negative binomial" was the key to much worth knowing about. As in Hollywood, a new name would be needed, or else nothing. On the other hand, the names power or scaling laws, by accident or design, suggest the stuff of legend.

6. My haphazard sampling of a large literature scattered across time and discipline suggests a conjecture: as the exponent in the power law approaches 1, the proponent approaches single-mindedness.

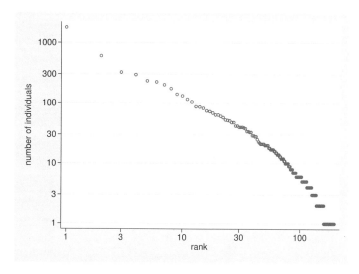

Figure 7: Power law fitted to macrolepidoptera data, Rothamsted 1935. The fit is good by some standards, but systematic curvature is also evident.

Once more, we need to consider alternatives but will go no further than the lognormal. It fits very well, except in the far tail of very rare species, but here the discreteness of the variable really causes problems (figure 8).

```
. qplot count, reverse trscale(invnorm(@)) xti(standard normal scale) ysc(log)
> yti(number of individuals) yla(1 3 10 30 100 300 1000, ang(h))
```

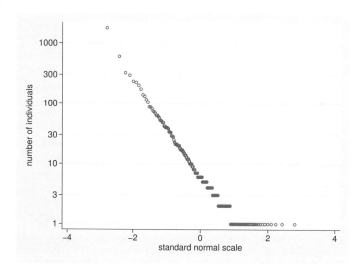

Figure 8: Lognormal distributions fitted to macrolepidoptera data, Rothamsted 1935. The fit is good except for species with counts near 1.

6   A Tukeyish aside

The discussion has moved from showing observed distributions, represented as quantile traces, to comparing those with fitted distributions, represented in the same way. Naturally, this is far from the only way to make the comparison. One alternative is a quantile–quantile plot, namely a scatter plot of observed and fitted quantiles.

One disadvantage of quantile plots used in this way is that both observed and fitted traces are necessarily monotone. In a sense, this can make agreement appear better than it really is, and one challenge is therefore to reduce or at least to remove the sloping behavior and make traces more nearly horizontal. This principle has been discussed in a previous column (Cox 2004b).

The particular case of power laws, or rather of very skewed so-called J-shaped distributions, is of interest because other graphical methods have been suggested for this situation, although they appear to have received little attention. Tukey (1977, chapter 18) noted a vague symmetry between two counts, the number of appearances and the number of individuals appearing at least that many times (the rank), and suggested the need for an analysis that treated these counts symmetrically. Given his attitude that square roots are the "first-aid" transformation for counts, Tukey thus proposed a plot of

$$\sqrt{\text{count} \times \text{rank}} \text{ versus } \log(\text{count}/\text{rank})$$

which he called a product-ratio plot. It is immediate that data for which

$$\text{count} \times \text{rank} = \text{constant}$$

follow a horizontal line on such a plot. An alternative in the same spirit as Tukey's proposal is a plot of

$$\log(\text{count} \times \text{rank}) \text{ versus } \log(\text{count}/\text{rank})$$

for which the same property holds. Thus power laws became more nearly horizontal. The idea can be implemented in a few lines:

```
. egen rank = rank(-count), unique
. generate y = log(count) + log(rank)
. generate x = log(count) - log(rank)
. label variable y "log(count X rank)"
. label variable x "log(count/rank)"
. scatter y x, xsc(reverse)
```

Although no results are shown here, this can be a helpful way of looking in finer detail at the structure of such distributions. How to handle tied counts is a small issue. Tukey (1977) suggested one procedure: the code above is based on another, which at least shows the frequency of ties fairly explicitly.

See also Parunak (1979) for other ideas on the same problem.

7   Distribution plots

The plot of cumulative probability versus ordered values we call here a distribution plot. Workers differ on plotting either the cumulative distribution or its complement, especially in the guise of a survival function. Official Stata offers many graphical and other tools for showing and analyzing the survival function for survival time data. It stops short at [R] **cumul**, which just calculates the usual cumulative distribution function, the subsequent graphics being a trivial application of whatever `twoway` plottype is desired. Nevertheless for convenience the `distplot` program has been offered (Cox 1999a), especially to make it easier to show several distributions simultaneously.

Broadly speaking, distribution plots and quantile plots have different bases. The first is based on rank/sample size as the estimate of cumulative probability. For some constant a, the second is based on so-called plotting positions defined by $(\text{rank} - a)/(\text{sample size} - 2a + 1)$: note that this definition does not include rank/sample size as a special case. The difference should not be dismissed as a mere nuance. With the latter convention, estimated cumulative probabilities will not usually be zero or one, and thus they will yield determinate results with some transformations, notably the logarithm and the logit. (The user is at liberty to insist that $a = 1$, however.) `distplot` has a `midpoint` option designed to reduce the number of zeros and ones, so that transformations as far as possible still apply, but the difference remains.

The difference between distribution plots and quantile plots is otherwise essentially one of choice of axes, which may seem to boil down to a question of habit or taste. However, Mandelbrot (1997, 207) makes a simple but valuable point: plots are not neutral, so different presentations of a dataset emphasize different aspects. Hence, for example, when the values that matter most are the largest, they are seen best in rank-size plots.

8   Software updates

This is a more formal statement flagging the release of new releases of `qplot` (gr42_3) and `distplot` (gr41_3) with this column. Previous versions of these programs appeared in Cox (1999b, 2001, 2004a) (`qplot`) and Cox (1999a, 2003a,b) (`distplot`). The main changes in this release are

1. Simplification of the main syntax. Previously, command calls would specify plottype as a subcommand, as in `qplot scatter` or `distplot line`. Now the plottype is by default as just exemplified and is otherwise changeable by use of `recast()`. See [G-3] *advanced_options*.

2. Previously the `by()` option specified that two or more groups be summarized by quantile curves within an individual graph panel. This was out of step with current Stata practice, so the `by()` option has been renamed `over()`. A true `by()` option, specifying use of different panels, has now also been added.

9 Conclusions

Ecological examples have been used to underline the flexibility of quantile plots, going beyond the possibilities of `quantile` and the defaults of `qplot`. Heavy use was made of the options of `qplot` that produce reversed, ranked, or transformed scales. It can be useful to take logarithms of quantiles, logarithms of ranks, or both. It is possible to carry out transformations of the probability or rank axis on the fly. Juxtaposed and superimposed comparisons are both straightforward. Quantile plots are indeed protean.

10 Acknowledgment

I thank Sarah A. Corbet for information on her father, A. Steven Corbet.

11 References

Bak, P. 1997. *How Nature Works: The Science of Self-Organized Criticality*. Oxford: Oxford University Press.

Batten, L. A. 1976. Bird communities of some Killarney woodlands. *Proceedings, Royal Irish Academy, Series B* 76: 285–313.

Bulmer, M. G. 1974. On fitting the Poisson lognormal distribution of species abundance data. *Biometrics* 30: 101–110.

Chambers, J. M., W. S. Cleveland, B. Kleiner, and P. A. Tukey. 1983. *Graphical Methods for Data Analysis*. Belmont, CA: Wadsworth.

Cleveland, W. S. 1993. *Visualizing Data*. Summit, NJ: Hobart.

———. 1994. *The Elements of Graphing Data*. Rev. ed. Summit, NJ: Hobart.

Cohen, J. E. 1966. *A Model of Simple Competition*. Cambridge, MA: Harvard University Press.

Cox, N. J. 1999a. gr41: Distribution function plots. *Stata Technical Bulletin* 51: 12–16. Reprinted in *Stata Technical Bulletin Reprints*, vol. 9, pp. 108–112. College Station, TX: Stata Press.

———. 1999b. gr42: Quantile plots, generalized. *Stata Technical Bulletin* 51: 16–18. Reprinted in *Stata Technical Bulletin Reprints*, vol. 9, pp. 113–116. College Station, TX: Stata Press.

———. 2001. gr42.1: Quantile plots, generalized: update to Stata 7.0. *Stata Technical Bulletin* 61: 10–11. Reprinted in *Stata Technical Bulletin Reprints*, vol. 10, pp. 55–56. College Station, TX: Stata Press.

———. 2002. Speaking Stata: How to move step by: step. *Stata Journal* 2: 86–102.

———. 2003a. Software update: gr41_1: Distribution function plots. *Stata Journal* 3: 211.

———. 2003b. Software update: gr41_2: Distribution function plots. *Stata Journal* 3: 449.

———. 2004a. Software update: gr42_2: Quantile plots, generalized. *Stata Journal* 4: 97.

———. 2004b. Speaking Stata: Graphing agreement and disagreement. *Stata Journal* 4: 329–349.

———. 2004c. Speaking Stata: Graphing distributions. *Stata Journal* 4: 66–88.

———. 2005. Speaking Stata: Density probability plots. *Stata Journal* 5: 259–273.

David, H. A., and H. N. Nagaraja. 2003. *Order Statistics*. 3rd ed. Hoboken, NJ: Wiley.

Domb, C. 1990. On Hammersley's method for one-dimensional covering problems. In *Disorder in Physical Systems*, ed. G. R. Grimmett and D. J. A. Welsh, 33–53. Oxford: Oxford University Press.

Fisher, R. A., A. S. Corbet, and C. B. Williams. 1943. The relation between the number of species and the number of individuals in a random sample of an animal population. *Journal of Animal Ecology* 12: 42–58.

Gordon, N. D., T. A. McMahon, B. L. Finlayson, C. J. Gippel, and R. J. Nathan. 2004. *Stream Hydrology: An Introduction for Ecologists*. Chichester, UK: Wiley.

Hutchinson, G. E. 1978. *An Introduction to Population Ecology*. New Haven, CT: Yale University Press.

Irwin, J. O. 1967. William Allen Whitworth and a hundred years of probability. *Journal of the Royal Statistical Society, Series A* 130: 147–176.

Kleiber, C., and S. Kotz. 2003. *Statistical Size Distributions in Economics and Actuarial Sciences*. Hoboken, NJ: Wiley.

Krebs, C. J. 1989. *Ecological Methodology*. New York: Harper and Row.

Lande, R., S. Engen, and B.-E. Sæther. 2003. *Stochastic Population Dynamics in Ecology and Conservation*. Oxford: Oxford University Press.

Magurran, A. E. 2004. *Measuring Biological Diversity*. Malden, MA: Blackwell.

Mandelbrot, B. B. 1997. *Fractals and Scaling in Finance: Discontinuity, Concentration, Risk*. New York: Springer.

Mosteller, F. 1965. *Fifty Challenging Problems in Probability with Solutions*. Reading, MA: Addison–Wesley.

Parunak, A. 1979. Graphical analysis of ranked counts (of words). *Journal of the American Statistical Association* 74: 25–30.

Perline, R. 2005. Strong, weak, and false inverse power laws. *Statistical Science* 20: 68–88.

Riley, N. D. 1948. Dr. A. S. Corbet. *Nature* 161: 1003.

Simon, H. A. 1991. *Models of My Life*. New York: Basic Books.

Tukey, J. W. 1977. *Exploratory Data Analysis*. Reading, MA: Addison–Wesley.

Whitworth, W. A. 1905. *Choice and Chance with One Thousand Exercises*. Facsimile edition, New York: Hafner, 1951. Previous editions 1867, 1870, 1878, 1886.

Williams, C. B. 1964. *Patterns in the Balance of Nature and Related Problems in Quantitative Ecology*. London: Academic Press.

Zipf, G. K. 1949. *Human Behavior and the Principle of Least Effort*. Cambridge, MA: Addison–Wesley.

About the author

Nicholas Cox is a statistically minded geographer at Durham University. He contributes talks, postings, FAQs, and programs to the Stata user community. He has also co-authored fifteen commands in official Stata. He was an author of several inserts in the *Stata Technical Bulletin* and is an Editor of the *Stata Journal*.

The Stata Journal (2005)
5, Number 4, pp. 574–593

Speaking Stata: Smoothing in various directions

Nicholas J. Cox
Durham University, UK
n.j.cox@durham.ac.uk

Abstract. Identifying patterns in bivariate data on a scatterplot remains a basic statistical problem, with special flavor when both variables are on the same footing. Ideas of double, diagonal, and polar smoothing inspired by Cleveland and McGill's 1984 paper in the *Journal of the American Statistical Association* are revisited with various examples from environmental datasets. Double smoothing means smoothing both y given x and x given y. Diagonal smoothing means smoothing based on the sum and difference of y and x that treats the two variables symmetrically, possibly under standardization. Polar smoothing is based on the transformation from Cartesian to polar coordinates followed by smoothing and then reverse transformation; here the smoothing is implemented by regression on a series of sine and cosine terms. These methods thus offer exploratory tools for determining the broad structure of bivariate data.

Keywords: gr0021, exploratory data analysis, statistical graphics, bivariate data, double smoothing, doublesm, diagonal smoothing, diagsm, polar smoothing, polarsm

1 Introduction

In my last two columns (Cox 2005a,b), I discussed some personal favorite methods that in some sense or another have failed to become totally standard. This theme continues with an examination of various ideas for smoothing data on scatterplots that were prominent in a major paper in a leading journal several years ago (Cleveland and McGill 1984) but which have not been widely adopted.

Smoothing has moved to center stage in statistical science over the last few decades. Not so long ago, smoothing meant, mostly, moving averages applied to time series. Moving averages are often useful, and they are tied to deep and beautiful mathematics in the frequency domain, but even in time-series analysis they are hardly the main idea. Yet in various ways, the search for smooth structures has become the focus of a large part of data analysis. This shift, as is well known, reflects new theory, new methods, and better computing facilities. It also reflects greater emphasis on identifying structures that both grow out of and can contribute to scientific explanations.

The interest in smoothing has grown hand in hand with a wider realization that regression, interpreted suitably broadly, is the most powerful single idea in statistics. Regardless of whether you agree with that, the theme of smoothing is broad enough to stretch from simple exploratory data analysis to deeper and more formally analyzed models. Much interest in smoothing begins with a relatively simple but perpetually interesting case, bivariate data on a scatterplot. It may seem surprising that anything can be said about that case that has not been said many times before, but this column surveys just one set of neglected ideas by way of example: double smoothing, diagonal smoothing, and polar smoothing. They are not competing methods, and they may even be complementary.

A few broad comments may be added at the outset on the homespun philosophy underlying this work. Scatterplots containing just data points do most of the work in conveying the information wrapped up in the two variables being examined. At most, we are adding enhancements that highlight some of the structure in the data—or highlight a lack of such structure. The question is often how far such enhancements support, or even contradict, any interpretation based mostly on "eyeballing" the data. In practice, such interpretations are also influenced by the scientific knowledge and the personal biases of the observer and are thus subjective to some extent. How far the data show what we think they show is a permanent issue.

2   Double and diagonal smoothing

Suppose that we have two variables that have the same units or are on equivalent unit-free scales (for example, they are both probabilities). This restriction can be relaxed, as will become clear. In addition, it is supposed that both are measured with error.

Examples are

1. Two sets of measurements of some quantity using different methods. Comparison of methods is a common problem throughout science and a much discussed subject in statistics (e.g., Dunn 2004).

2. Two sets of measurements at nearby locations. Concretely, we might have precipitation or temperature data from nearby meteorological stations.

3. Observed responses and predictions from some model. This model need not have a statistical flavor.

4. Corresponding quantiles for two groups of a variable or two variables, as might be shown on a quantile–quantile plot.

In these examples, and in many others, equality of variables is a natural reference so that equality is either expected or at least a benchmark.

If we show data for these variables on a scatterplot, which variable goes on the y-axis and which on the x-axis is likely to be arbitrary. The examples are all variables that are essentially on the same footing. We might, however, want to add some smooth trace to the graph to help to identify any systematic pattern. The motivation here is exploratory or heuristic. We might prefer not to specify formally what error structure is associated with the variables, marginally or jointly.

This problem is therefore some distance from the classic smoothing problem in which one variable fluctuates erratically and the other variable (for example, time) is considered known. The problem of "errors in variables" is related but not identical. Evidently, neither smoothing y given x nor smoothing x given y would be entirely suitable. We could

1. Choose one of these, say, smoothing y given x. At best, if the relationship is strong, the other smooth will be very similar. This is in principle wrong, as it assumes that one variable is known exactly, although we might choose not to worry about that.

2. Choose and plot both. Although not often done, this could be interesting whatever was found. Either the traces are similar—so the choice is in fact not an issue—or they are different, which should provoke further thought. This practice I here call *double smoothing*. My inspiration is Cleveland and McGill (1984, 817–818).

3. Choose a smoothing method that treats y and x symmetrically. This is even less often done, but again an ingeniously simple idea from Cleveland and McGill (1984, 818) offers a way forward. They call their procedure *sum-difference smoothing*, but I propose the name *diagonal smoothing* to reflect the idea that smoothing is done diagonally, rather than parallel to one of the axes.

Whenever x and y are positively correlated, the procedure is to form the difference and sum

$$d = y - x \qquad\qquad s = y + x$$

and then smooth the difference as a function of sum in some way, calling the result $\widehat{d}$. The calculation is then reversed, producing

$$\widehat{y} = (s + \widehat{d}\,)/2 \qquad\qquad \widehat{x} = (s - \widehat{d}\,)/2$$

to plot with the data points y and x.

Whenever x and y are negatively correlated, the procedure is to form the difference and sum, as before, and then smooth the sum as a function of the difference in some way, calling the result $\widehat{s}$. The calculation is then reversed, producing

$$\widehat{y} = (d + \widehat{s})/2 \qquad\qquad \widehat{x} = (\widehat{s} - d)/2$$

to plot with the data points y and x. (The case of negative correlation is not discussed by Cleveland and McGill but is in practice much less common in problems in which this technique is most appealing.)

This method does not depend on $y - x$ being roughly constant, but it should work well whenever that is so. But calculating the difference and sum does depend on y and x sharing the same units. However, it is easy to ensure comparability by prior use of some method of standardization, followed by smoothing on standardized scales and a reversal of the standardization.

2.1 Double smoothing in Stata

Double smoothing in Stata is straightforward. In practice, a wrapper program to do it is helpful. One is published with this column as `doublesm`. Nevertheless, let us consider how to attack the problem. Suppose that `lowess` is the smoothing command of choice and our variables are y and x. The first step is to smooth y on x and keep the smoothed values:

```
lowess y x, nograph gen(ys)
```

The second step is to smooth x on y and keep the smoothed values:

```
lowess x y, nograph gen(xs)
```

Then we put together the results in a graph. The main point requiring care is the sort order for each graph component:

```
twoway line ys x, sort || ///
       line y xs, sort(y xs) || ///
       scatter y x
```

Naturally, this schematic outline omits details of how the `lowess` calls and, in turn, the graph call might be tuned using various options. But a wrinkle does deserve mention: with `lowess` as implemented in Stata, the same smooth is not guaranteed for the same (x, y) pairs. Normally this is not a great problem. If a graphical display makes nonuniqueness evident, it is just drawing attention to a genuine characteristic of the data, the occurrence of ties. However, it is worth noting that this may happen.

With `doublesm`, this strategy is implemented via a single call `doublesm y x`; smoothing using `lowess` is the default. Optionally, the running line method as implemented by Sasieni (1995); Sasieni and Royston (1998); and Sasieni, Royston, and Cox (2005) may be specified. Earnest programmers wishing to add their favorite smoothing method would find it easy to adapt `doublesm`. In general, any method of smoothing that copes with unequal spacing of values is likely to be adequate for exploration. The usual comment from both theory and experience is that the degree of smoothing is more important than the particular flavor implemented.

2.2 Diagonal smoothing in Stata

As before, the main question in a Stata implementation of diagonal smoothing is precisely how to smooth. A program `diagsm` published with this column uses [R] **lowess** by default and running line smoothing optionally.

`diagsm` by default standardizes each variable using (value − median) / MAD before smoothing. Here MAD is the median absolute deviation from the median. This should be a reasonable general choice, given the possibility of variables with skewed and/or heavy-tailed distributions. Note, however, that MAD may be 0 whenever half or more of the values are equal to the median, in which circumstance the method will not work. A real example is the repair record variable rep78 in Stata's auto dataset for domestic cars.

A less robust standardization available through the meansd option is (value − mean)/ (standard deviation). In each case, standardization is reversed to produce smoothed values. Only by specification of the raw option is smoothing on the original data scales available. This may seems backwards design, given the motivation of the method, but the intent is to protect naive users from the effects of what may well be an inappropriate default.

3 Polar smoothing

One idea that several people have tried to transfer from the univariate case to the bivariate is that of the middle or central part of the data. The ideas of the median as a center and of the quartiles as defining the central half of a distribution, from an order statistic point of view, were very well established by the latter nineteenth century, most prominently in the work of Francis Galton. Naturally, the middle or central part— whether precisely or roughly one half of the data—defines as a complement the outer part, which may well be practically or scientifically even more important or interesting.

There are various possibilities for extending these ideas to two variables and no apparent reason why several should not be fruitful. One is based on the idea of convex hulls (e.g., Barnett 1976 and various papers in Barnett 1981). Others are based closely on the box plot, generalized in some fashion (e.g., Goldberg and Iglewicz 1992 and Rousseeuw, Ruts, and Tukey 1999). Here we focus on what Cleveland and McGill (1984, 819–820) called polar smoothing. They cite a 1977 personal communication from Alan M. Gross.

Given data on x and y, the main idea is to map to polar coordinates, smooth the radius with respect to its angle, and then map back again. The smoothing thus defines a closed curve including the central part of the data. Including half of the values is not guaranteed but is often approximately satisfied. Generalizing mildly from the formulation of Cleveland and McGill, one algorithm runs like this:

1. Given measures of location $m(x)$ and $m(y)$ and measures of scale (width) $w(x)$ and $w(y)$, form $x' = \{x - m(x)\}/w(x)$, $y' = \{y - m(y)\}/w(y)$.

2. Convert these to normalized coordinates $d = y' - x'$, $s = y' + x'$.

3. Normalize in turn to $d' = d/w(d)$, $s' = s/w(s)$.

4. Convert to polar coordinates $r = \sqrt{s'^2 + d'^2}$, $\theta = \arctan(d/s)$.

5. Transform r to $z = r^{2/3}$ to provide some robustness, thus pulling in the larger spikes particularly. As r is a square root, this produces a cube root: compare

the Wilson–Hilferty transform that arises particularly in treatments of gamma distributions.

6. Smooth z as a function of θ, producing $\widehat{z}$.

7. Reverse 5 by $\widehat{r} = \widehat{z}^{3/2}$.

8. Reverse 4 by $\widehat{d'} = \widehat{r}\sin\theta$, $\widehat{s'} = \widehat{r}\cos\theta$.

9. Reverse 3 to get $\widehat{d} = w(d)\,\widehat{d'}$, $\widehat{s} = w(s)\,\widehat{s'}$.

10. Reverse 2 to get $\widehat{x'} = (\widehat{s} - \widehat{d})/2$, $\widehat{y'} = (\widehat{s} + \widehat{d})/2$.

11. Reverse 1 to get $\widehat{x} = m(x) + w(x)\,\widehat{x'}$, $\widehat{y} = m(y) + w(y)\,\widehat{y'}$.

What is tunable here?

Measures of location and scale. Cleveland and McGill use median for location and MAD, namely, median absolute deviation from the median, as measure of scale. Other possibilities are mean for location and standard deviation for scale. Yet another set of possibilities is not to normalize or equivalently to set $w(x) = w(y) = 1$.

Robustness step. The cube root transform could be varied or even omitted. But without some "robustifying" step, the representation of the central part would inevitably be sensitive to outliers, which is arguably contrary to the main purpose here.

Smoother. The most interesting issue is how to smooth the radii of polar coordinates. It is crucial to ensure that the smoother wraps around the circle without jumps or kinks and so respects the fact of working with respect to polar coordinates. Cleveland and McGill copy data fore and aft of the range of 360° or 2π radians and then apply lowess, and many other smoothers could be applied in the same way.

3.1   Polar smoothing in Stata

A program `polarsm` is published with this column. I have found it easiest to smooth by means of a regression on pairs of sine and cosine terms. Thus we `regress` z on $\sin j\theta$, $\cos j\theta$ for $j = 1$ to whatever k is desired. As sine and cosine functions automatically wrap around the circle, no special trickery of copying fore and aft is needed.[1] If $k = 1$, then some $b_0 + b_1\sin\theta + b_2\cos\theta$ is fitted to the transformed polar coordinates. This expression has a single maximum and a single minimum half a circle away from that. The effect when back-transformed to Cartesian coordinates thus resembles fitting an ellipse that represents the central part of the data. More pairs of terms (a larger k) allow more complication, and the resulting curve can then display concave as well as convex parts. The choice of k can, and arguably should, be informal; use whatever

---

1. So, the slogan is not trickery, but trigonometry! Not fudge, but Fourier!

seems to work best at revealing structure in the data. If you have too high a value of
k, the curve tends towards an interpolation of the data, yielding a slightly spectacular
but utterly useless starburst pattern. The only common-sense stipulation is to smooth
groups to be compared to similar degrees.

Mentioning groups leads to one of the most interesting ways to apply polar smoothing, either to separate y variables or to separate groups defined by some third variable
not shown on the scatterplot. polarsm allows either of these choices. In the latter case,
groups are indicated by an over() option. Either way, the question then is how far the
polar smooths differ or are similar.

By default, polar smooths are shown superimposed, but you may also wish to show
them in separate panels. This can be done by specifying by(), as usual. If separate
y variables are being shown, know that polarsm temporarily restructures the dataset
using stack. The option by(_stack) is thus the way to produce separate panels. Any
preexisting _stack is temporarily dropped from the data. If separate groups are indicated by over(*groupvar*), using an additional by(*groupvar*) is the way.

Although the polar smooth is, by its definition as a sum of sines and cosines, a
very smooth function, it is shown by polarsm as a connected line. Strictly, it is shown
as two connected lines, one connecting the first to the last smoothed value, and the
other closing the loop by connecting the last to the first smoothed value, but they
share a linestyle. The reasons for using connected lines are both negative and positive.
Negatively, twoway mspline can show only single-valued curves, not loops. Positively,
reminding you that the smooth is an interpolation between distinct values, one for each
data point, might even be a small feature.

4   The doublesm, diagsm, and polarsm programs

4.1   Syntax for doublesm

doublesm *yvar* *xvar* [*if*] [*in*] [,

  [lowess(*lowess_options*) | running [(*running_options*)]] line(*line_options*)

  xsmooth(*line_options*) ysmooth(*line_options*) data(*scatter_options*)

  [addplot(*plot*) | plot(*plot*)] *graph_options*]

4.2   Syntax for diagsm

diagsm *yvar* *xvar* [*if*] [*in*] [, [raw | meansd]

  [lowess(*lowess_options*) | running [(*running_options*)]] line(*line_options*)

  data(*scatter_options*) [addplot(*plot*) | plot(*plot*)] generate(*newyvar*

  newxvar) *graph_options*]

4.3 Syntax for polarsm

polarsm *varlist* [*if*] [*in*] [, terms(*#*) <u>o</u>ver(*groupvar*)
   [colors(*colorstyle*) | colours(*colorstyle*)] smooth(*line_options*)
   data(*scatter_options*) addplot(*plot*) *graph_options*]

4.4 Options for doublesm

lowess(*lowess_options*) specifies options for lowess; see [R] **lowess**.

running[(*running_options*)] specifies the use of the command running as an alternative to lowess. The option running by itself specifies using the running command
   and its defaults. Alternatively, running() specified with arguments passes options
   to the running command.

line(*line_options*) specifies options for line (see [G-2] **graph twoway line**) that are
   to be applied to both smooth traces.

xsmooth(*line_options*) specifies options for line (see [G-2] **graph twoway line**) that
   are to be applied to the smooth of *xvar* given *yvar*.

ysmooth(*line_options*) specifies options for line (see [G-2] **graph twoway line**) that
   are to be applied to the smooth of *yvar* given *xvar*.

data(*scatter_options*) specifies options for scatter (see [G-2] **graph twoway scatter**)
   that are to be applied to rendering of the data points.

addplot(*plot*) (Stata 9 and later; see [G-3] ***addplot_option***) or plot(*plot*) (Stata 8
   only; see [G-3] ***plot_option***) provides a way to add other plots to the generated
   graph.

graph_options are other options documented in [G-3] ***twoway_options***.

4.5 Options for diagsm

raw specifies no scaling or standardization so that the sum and difference are calculated
   directly from the data before smoothing.

meansd specifies scaling by (value − mean) / standard deviation before smoothing. The
   standardization will be reversed after smoothing.

lowess(*lowess_options*) specifies options for lowess; see [R] **lowess**.

running[(*running_options*)] specifies the use of the command running as an alternative to lowess. The option running by itself specifies using the running command
   and its defaults. Alternatively, running() specified with arguments passes options
   to the running command.

line(*line_options*) specifies options for line (see [G-2] **graph twoway line**) that are
   to be applied to the smooth trace.

data(*scatter_options*) specifies options for scatter (see [G-2] **graph twoway scatter**) that are to be applied to the data points.

addplot(*plot*) (Stata 9 and later; see [G-3] ***addplot_option***) or plot(*plot*) (Stata 8 only; see [G-3] ***plot_option***) provides a way to add other plots to the generated graph.

generate(*newyvar newxvar*) specifies the names of two new variables to hold the y and x coordinates of the diagonal smooth.

graph_options are other options documented in [G-3] ***twoway_options***.

4.6 Options for polarsm

terms(*#*) specifies the number k of Fourier terms, $\sin j\theta$, $\cos j\theta$, $j = 1, \ldots, k$ on which the radius is regressed as a function of the polar angle θ. By default, k is 1.

over(*groupvar*) specifies a variable to be used to identify different groups. over() is not allowed whenever three or more variables are specified.

colors(*colorstyle*) or colours(*colorstyle*) (choose just one) specifies a list of colors with which to show different polar smooths. See [G-3] ***colorstyle***.

smooth(*line_options*) specifies options for line (see [G-2] **graph twoway line**) controlling the representation of different polar smooths.

data(*scatter_options*) specifies options for scatter (see [G-2] **graph twoway scatter**) controlling the representation of data points for different variables or subsets of data.

addplot(*plot*) provides a way to add other plots to the generated graph; see [G-3] ***addplot_option***.

graph_options are other options documented in [G-3] ***twoway_options***. In particular, note that by default, polar smooths are shown superimposed. You may want to show them in separate panels by specifying by(). If separate y variables are being shown, polarsm temporarily restructures the dataset using stack; see [D] **stack**. The option by(_stack) is thus the way to produce separate panels. (Any preexisting _stack is temporarily dropped from the data.) If separate groups are specified by over(*groupvar*), an additional by(*groupvar*) is the way to go.

5 Examples

Two substantial examples using environmental datasets will now be examined. Polar smoothing also features briefly in the engaging text by Helsel and Hirsch (1992).

5.1   Rainfall in the southern Pennines

A first example features the rainfall captured at three rain gauges over 42 years (1906–1947) in part of the Derwent catchment (drainage basin or watershed) in the southern Pennines in Britain. Rainfall means, strictly, precipitation: any fall of snow or other solid forms of water is included as rainfall equivalent. The original data as reported by Law (1953) were given in inches, which are long since obsolete as measurement units, except among historians and U.S. citizens. However, those units are retained here: for your information, 1 inch is exactly 25.4 mm.

As rainfall is sensitive to small differences in geographic location, especially but not only landsurface altitude, many rain gauges are needed to get a good idea of the total or average rainfall in an area. Conversely, rainfalls at nearby places do tend to be similar. A common hydrological practice, as rain gauges are expensive, is thus to pepper an area initially with many gauges, and then find out which pairs of gauges correlate so highly that one can be removed without much loss of information. Standard climatological folklore runs (e.g., Linacre 1992, Dingman 2002) that ratios of rainfalls at nearby rain gauges tend to be nearly constant, but naturally this needs checking. Even if it holds, the constant needs to be determined from data.

What should influence analysis is that the amounts at different gauges are variables on the same footing. It would be at best arbitrary, and at worst wrong, to regard the rainfall *here* as a response to the rainfall *there*, or indeed vice versa. It would be nearer the mark to regard both as related to some latent or hidden variable, but here we proceed otherwise, merely treating pairs of variables symmetrically.

Let us start with the idea of double smoothing, the idea of smoothing x given y, as well as that of smoothing y given x. Figures 1, 2, and 3 show the results of double smoothing pairs of three gauges, numbered 17, 23, and 60. More generally, double smoothing pairs of k variables would yield $k(k-1)/2$ plots, taking interchange of axes for granted, just as with a half of a scatterplot matrix. The Stata recipe for each of these graphs resembles this command:

```
. doublesm srain23 srain17, yla(, nogrid ang(h)) ms(oh)
> legend(order(1 "23 | 17" 2 "17 | 23") ring(0) pos(5) col(1))
> addplot(function x, range(srain23) lstyle(grid))
```

As the different gauges are highly correlated, it is possible to tuck the legend into the bottom right-hand corner, although taste or convention might lead you to another choice. A reference line $y = x$ is also added. This could be done in various ways, but putting a `twoway function` call into an `addplot()` option and requesting a `grid` linestyle is one. (In Stata 8, use a `plot()` option instead.)

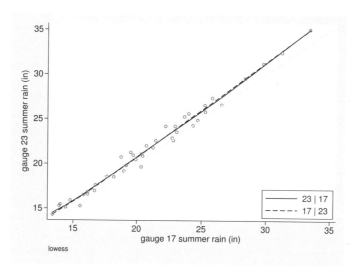

Figure 1: Double smoothing by lowess of summer rainfall at two rain gauges. The smooths of gauge 23 given 17 and of gauge 17 given 23 are almost identical and shifted slightly from the reference line $y = x$.

The graphs show part of the range of behavior that might be expected. When two variables agree quite closely, as with gauges 17 and 23 in figure 1, the two smooths are essentially the same, and both are close to the reference line of equality. The structure in this example seems more additive than multiplicative, although there is a hint of multiplicativity at higher rainfalls.

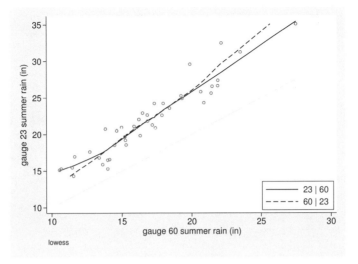

Figure 2: Double smoothing by lowess of summer rainfall at two rain gauges. The smooths of gauge 23 given 60 and of gauge 60 given 23 are similar. Both indicate a mainly multiplicative shift.

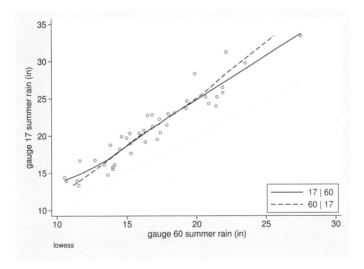

Figure 3: Double smoothing by lowess of summer rainfall at two rain gauges. The smooths of gauge 17 given 60 and of gauge 60 given 17 are similar. Again both indicate a mainly multiplicative shift.

When pairs of variables do not agree so closely, the smooths inevitably differ to some extent (figures 2 and 3). What is most common is a scissors pattern of intersecting smooths, familiar from any discussion of the two regression lines of y given x and of

x given y. In either case, however, the smooths in these two figures both support the notion of multiplicative structure. This could be explored in other ways, such as plotting the difference versus the mean or the ratio versus the geometric mean, an approach discussed, with several references, in a previous column (Cox 2004a).

When the double smooths agree, many would be happy to accept a compromise between them, and this is where diagonal smoothing is most congenial. Figure 4 gives one example. A generate() option is used to store the coordinates of the smooth for later use. As earlier mentioned, the default of diagsm is to smooth after a robust standardization, which is then reversed.

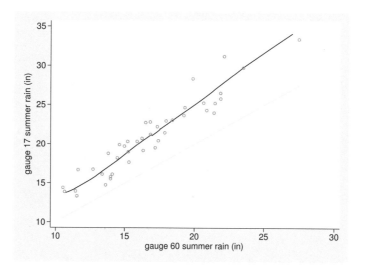

Figure 4: Diagonal smoothing by lowess of summer rainfall at two rain gauges, 17 and 60.

```
. diagsm srain17 srain60, yla(, nogrid ang(h)) ms(oh)
> legend(ring(0) pos(5) col(1)) addplot(function x, range(srain60) lstyle(grid))
> generate(y_1760 x_1760)
```

The new variables can then be added to a composite graph (figure 5).

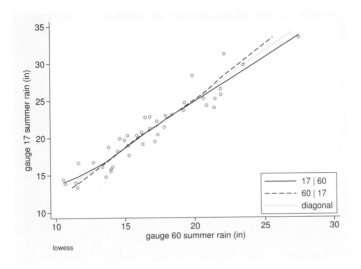

Figure 5: Double and diagonal smoothing by lowess of summer rainfall at two rain gauges, 17 and 60. The diagonal smooth provides a reasonable compromise between the double smooths.

```
. doublesm srain17 srain60, yla(, nogrid ang(h)) ms(oh)
> addplot(function x, range(srain60) lstyle(grid) ||
> line y_1760 x_1760, lp(dot) sort)
> legend(order(1 "17 | 60" 2 "60 | 17" 5 "diagonal") ring(0) pos(5) col(1))
```

The `legend(order())` call shows the benefit of insider information. Know that the fifth variable being plotted here is the diagonal smooth: the third is the data, as a set of point symbols, and the fourth is the line of equality.

This graph is, as you will appreciate, a well-behaved example, and experience will differ with worse behavior. Whenever candidate smooths agree, it matters little, either scientifically or even statistically, which one we choose, but at least in this case we considered different possibilities for analysis. Even in worse cases, differences between candidate smooths will emphasize uncertainty about how best to summarize the structure, and thus caution against overhasty interpretation.

5.2 Cirques in the English Lake District

We continue with a geomorphological example, returning to a dataset used previously in these columns (Cox 2004c, 2005a) on 158 glacial cirques from the English Lake District (Evans and Cox 1995). Glacial cirques are hollows excavated by glaciers that are open downstream, bounded upstream by the crest of a steep slope, and arcuate in plan around a more gently sloping floor. More informally, they are sometimes described as 'armchair-shaped'. A similar dataset from Wales is used elsewhere in this issue (Cox 2005c).

Two key variables are the length and width of these cirques. Previous analyses indicate on various grounds that it is best to think on logarithmic scales, so we smooth on those scales. However, we will want graphs to be labeled in terms of the original variable labels and units, which make much more scientific sense. This is a common-enough desire in other problems for us to detail one way how to do that.

First, set up default titles and axis labels in local macros. The program `mylabels` is downloadable from SSC and was exemplified in an earlier column (Cox 2004b).

```
. local yttl "Width, max. perpendicular to median axis (m)"
. local xttl "Length of median axis, focus to crest (m)"
. mylabels 200 500 1000 2000, myscale(log10(@)) local(la)
```

Width and length are, once again, variables on the same footing. Geometrically, both are aspects of plan size and shape. Physically, neither is a forcing variable. So diagonal and double smoothing appeal here. Figure 6 was produced similarly to figure 5 by first diagonal smoothing, using its `generate()` option to save the coordinates of the smooth, and then plotting those on top of a double smooth. What is especially intriguing and a new detail is the hint of curvature at high values. Power-function models, which transform to straight lines when both variables are logged, have long been popular with variables like these, but the graph hints that here the power function may not fit quite as well as thought. Other analyses are needed to follow this up, but the smoothing has produced a new idea.

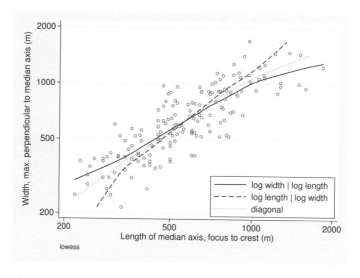

Figure 6: Double and diagonal smoothing by lowess of log cirque width and length. Smoothing was carried out on logarithmic scales. There is a hint of curvature at high values that conflicts slightly with the idea of a power function fit, which would imply a straight line fit on this graph.

Now we turn to examples of polar smoothing. In practice, start with the default in which one pair of terms (one sine and one cosine) is used to smooth polar coordinates, thus yielding an approximate ellipse (figure 7). Then add further terms until it seems that they are merely corresponding to noise or uninteresting detail (figures 8 and 9).

```
. polarsm logw logl, yla(`la´, ang(h)) xla(`la´) yti("`yttl´") xti("`xttl´") ms(oh)
. polarsm logw logl, yla(`la´, ang(h)) xla(`la´) yti("`yttl´") xti("`xttl´") ms(oh)
> terms(2)
. polarsm logw logl, yla(`la´, ang(h)) xla(`la´) yti("`yttl´") xti("`xttl´") ms(oh)
> terms(4)
```

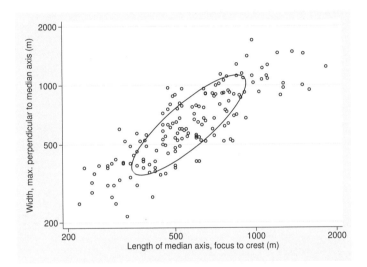

Figure 7: Polar smooth of log cirque width and length. Smoothing was carried out on logarithmic data. The smoothing is based on regressing radial coordinates on one sine term and one cosine term, producing an approximate ellipse on back-transformation.

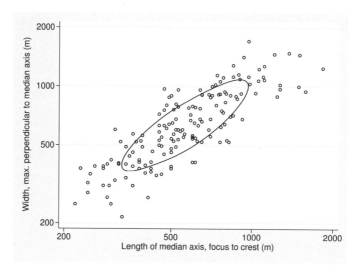

Figure 8: Polar smooth of log cirque width and length. Smoothing was carried out on logarithmic data. The smoothing is based on regressing radial coordinates on two sine terms and two cosine terms, but the result resembles figure 7.

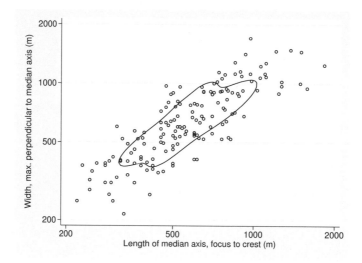

Figure 9: Polar smooth of log cirque width and length. Smoothing was carried out on logarithmic data. The smoothing is based on regressing radial coordinates on four sine terms and four cosine terms and shows some sensitivity to small details of the configuration of points.

As earlier mentioned, polar smoothing done groupwise is a way of checking whether groups behave similarly or differently. An ordered categorical variable in this dataset is the 'grade' of the cirque, codifying the expert's best summary of how well developed the cirque is, meaning how far it represents evidence of strong modification by ice. The numbers in each grade are classic 11, well-defined 31, definite 44, marginal 37, and poor 35. The number of classic cirques is too small for this technique, so the first two categories are combined:

```
. generate grade2 = cond(grade <= 2, 1, grade)
. label define grade2 1 "classic, well-defined" 3 "definite" 4 "poor"
> 5 "marginal"
. label values grade2 grade2
. polarsm logw logl, yla(`la´, ang(h)) xla(`la´) yti("`yttl´") xti("`xttl´")
> ms(oh) terms(2) over(grade2) by(grade2, row(1) legend(off) note("") compact)
> subtitle(, ring(0) pos(11) nobexpand box fcolor(none)) xla(, ang(v))
```

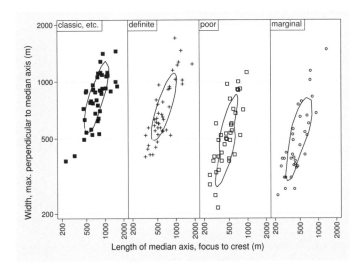

Figure 10: Polar smooth of log cirque width and length, separately by expert-determined grade. Smoothing was carried out on logarithmic data. A systematic shift over categories is evident, although results indicate that poor and marginal cirques are similar.

The polar smooths in figure 10 show evidence of a systematic change in characteristic lengths and widths but also some novel details, including the approximate similarity of the last two categories.

6   A note on quantile–quantile plots

An application of diagonal smoothing particularly that is not discussed in detail here is to quantile–quantile plots. These plots, for example, as implemented in Stata's `qqplot`, are often plainly presented, usually with no more than a reference line. Yet adding a smooth can be useful. For this, it is helpful to have a utility generating corresponding quantiles for two variables or for two groups of one variable. `cquantile` is built for this purpose and may be downloaded using `ssc` ([R] **ssc**). A quantile–quantile plot with smoothing can then be produced with `diagsm`.

7 Conclusions

Cleveland and McGill (1984) presented an abundance of simple but valuable ideas in their wide-ranging paper, perhaps too many for them all to be absorbed. Their 'sunflowers' have seeded and borne fruit recently in Stata (see [R] **sunflower**; Dupont and Plummer, Jr., 2005). This paper has drawn attention to an array of more neglected ideas for identifying basic structure on scatterplots, together with the release of new programs for experimentation, `doublesm`, `diagsm`, and `polarsm`. They are particularly relevant to the common but rather neglected problem of smoothing when both variables are on the same footing.

8 Acknowledgments

William S. Cleveland, Marcello Pagano, and Vincent Wiggins provided encouragement. Continuing joint work with Ian S. Evans provided data and motivation.

9 References

Barnett, V. 1976. The ordering of multivariate data. *Journal of the Royal Statistical Society, Series A* 139: 318–355.

Barnett, V., ed. 1981. *Interpreting Multivariate Data*. Chichester, UK: Wiley.

Cleveland, W. S., and R. McGill. 1984. The many faces of a scatterplot. *Journal of the American Statistical Association* 79: 807–822.

Cox, N. J. 2004a. Speaking Stata: Graphing agreement and disagreement. *Stata Journal* 4: 329–349.

―――. 2004b. Speaking Stata: Graphing categorical and compositional data. *Stata Journal* 4: 190–215.

―――. 2004c. Speaking Stata: Graphing distributions. *Stata Journal* 4: 66–88.

―――. 2005a. Speaking Stata: Density probability plots. *Stata Journal* 5: 259–273.

―――. 2005b. Speaking Stata: The protean quantile plot. *Stata Journal* 5: 442–460.

―――. 2005c. Stata tip 27: Classifying data points on scatter plots. *Stata Journal* 5: 604–606.

Dingman, S. L. 2002. *Physical Hydrology*. Upper Saddle River, NJ: Prentice Hall.

Dunn, G. 2004. *Statistical Evaluation of Measurement Errors: Design and Analysis of Reliability Studies*. London: Arnold.

Dupont, W. D., and W. D. Plummer, Jr. 2005. Using sunflower plots to explore bivariate relationships in dense data. *Stata Journal* 5: 371–384.

Evans, I. S., and N. J. Cox. 1995. The form of glacial cirques in the English Lake District, Cumbria. *Zeitschrift für Geomorphologie* 39: 175–202.

Goldberg, K., and B. Iglewicz. 1992. Bivariate extensions of the boxplot. *Technometrics* 34: 307–320.

Helsel, D. R., and R. M. Hirsch. 1992. *Statistical Methods in Water Resources*. Amsterdam: Elsevier. http://pubs.usgs.gov/twri/twri4a3/.

Law, F. 1953. The estimation of the reliable yield of a catchment by correlation of rainfall and run-off. *Journal of the Institution of Water Engineers* 7: 273–293.

Linacre, E. T. 1992. *Climate Data and Resources: A Reference and Guide*. London: Routledge.

Rousseeuw, P., I. Ruts, and J. W. Tukey. 1999. The bagplot, a bivariate boxplot. *American Statistician* 53: 382–387.

Sasieni, P. 1995. sed9: Symmetric nearest neighbor linear smoothers. *Stata Technical Bulletin* 24: 10–14. Reprinted in *Stata Technical Bulletin Reprints*, vol. 4, pp. 97–101. College Station, TX: Stata Press.

Sasieni, P., and P. Royston. 1998. sed9.1: Pointwise confidence intervals for running. *Stata Technical Bulletin* 41: 17–23. Reprinted in *Stata Technical Bulletin Reprints*, vol. 7, pp. 156–163. College Station, TX: Stata Press.

Sasieni, P., P. Royston, and N. J. Cox. 2005. sed9_2: Symmetric nearest neighbor linear smoothers. *Stata Journal* 5: 285.

About the author

Nicholas Cox is a statistically minded geographer at Durham University. He contributes talks, postings, FAQs, and programs to the Stata user community. He has also co-authored fifteen commands in official Stata. He was an author of several inserts in the *Stata Technical Bulletin* and is an Editor of the *Stata Journal*.

The Stata Journal (2006)
6, Number 3, pp. 397–419

Speaking Stata: Graphs for all seasons

Nicholas J. Cox
Department of Geography
Durham University
Durham City, UK
n.j.cox@durham.ac.uk

Abstract. Time series showing seasonality—marked variation with time of year—are of interest to many scientists, including climatologists, other environmental scientists, epidemiologists, and economists. The usual graphs plotting response variables against time, or even time of year, are not always the most effective at showing the fine structure of seasonality. I survey various modifications of the usual graphs and other kinds of graphs with a range of examples. Although I introduce here two new Stata commands, `cycleplot` and `sliceplot`, I emphasize exploiting standard functions, data management commands, and graph options to get the graphs desired.

Keywords: gr0025, cycleplot, sliceplot, seasonality, time series, graphics, cycle plot, rotation, state space, incidence plots, folding, repeating

1 Seasonality

Seasonality—marked variation with time of year—must have been evident to the first humans. Indeed many organisms show awareness of, or adaptations to, seasonality. It remains a matter of great interest to many scientists.

Astronomers explain seasonality in terms of the motion of the earth relative to the sun. That story is part of one of the great successes of modern science, which we owe largely to Copernicus, Kepler, and Newton. Viewed astronomically, seasonality—for example, prediction of times of sunrise or sunset—is a classic deterministic problem, but for all other sciences it has a strongly stochastic or statistical flavor. Climatologists look at variations in temperature, rainfall, and other elements around the year, but everyone knows that no two summers are identical. Seasonality of climate has many other environmental effects. Many are fairly direct, such as those on water supply or vegetation condition, but some are more subtle and even controversial, such as alleged seasonality in the incidence of earthquakes or volcanic eruptions in response to variations in overburden pressure. Epidemiologists examine seasonal variations in morbidity, mortality, and natality, an approach that goes back at least as far as the Hippocratic writing *Airs, Waters, Places* in the fifth century BCE. Economists have long monitored seasonal variations in variables such as employment, sales, and GDP, although often these are regarded as nuisances requiring seasonal adjustment.

The most common graphs for seasonal data are plots of one or more response variables versus time or time of year. This statement is surely well known, so why then this column? Negatively, such plots are often not especially effective at showing the fine structure of seasonality. Positively, their effectiveness can be improved by various tricks, and other kinds of plots can be useful too: indeed, we can borrow ideas on seasonal graphics from various fields.

I will introduce two user-written commands, `cycleplot` and `sliceplot`, but I will emphasize using some basic functions, graphics options, and data management commands.

This column is the second of a series with the general theme of circular arguments. The first column examined time of day as a circular scale (Cox 2006).

2   Related problems

Although the focus here is on seasonality, the main ideas carry over to other periodicities, such as time of day or time of week. I will not spell out that connection further, as translating code to other periodicities will typically be straightforward. Similarly, just flagging a standard point should be enough: seasonality is usually combined with variations on other time scales. The graphics to be discussed apply either to data with some seasonal variation or to a seasonally varying component of such data, calculated in some way.

Traditionally, we distinguish seasons by named divisions: in English, as winter, spring, summer, and autumn or fall. In climatology, these divisions are often made more precise as the four quarters December–February, March–May, June–August, and September–November, because surface phenomena tend to lag solar inputs enough to justify the offset of 1 month from the conventional calendar year beginning in January. In data analysis, any such divisions are usually at best conventional or convenient categories. Underlying them are periodic or circular numerical scales, such as month of year or day of year, in which the last value of any year is followed by the first value of the following year.

How far, then, should seasonal data be considered a kind of circular data? Some intriguing circular graphs have been suggested for seasonal data. For example, Tufte (2001, 72) reproduces a spiral representation of Italian postal bank deposits from 1876 to 1881. Unfortunately, reading off the structure of seasonality from such graphs is hard. I suggest that, on the whole, seasonal data are better shown using linear graphics. This conclusion follows partly because seasonal data are one kind of time series, for which a linear time axis is both customary and natural, and partly because few scientists have much experience in interpreting seasonal graphics displayed in circular formats, in contrast to their frequent familiarity with compass or map formats. Brinton (1914, 80) aired a similar view.

That said, one elementary but also fundamental idea is worth borrowing in seasonal graphics and has already been hinted at. January is an arbitrary start to the year in almost all senses but calendar convention, so rotating the seasons to start the time-of-year scale at another time may be useful. The concept is already familiar to those accustomed to thinking in financial or fiscal years.

The examples here are all for time series in the strict sense: variables counted or measured for regularly spaced times, whether intervals or points. There are also event data, times for deaths, earthquakes, riots, and so forth. Ideas for graphing the occurrence or frequency of such point process data follow readily from the ideas to be discussed here.

With its focus on graphics, this column cannot do justice to a theme that is linked but also distinct: how best to model (or smooth) time series, given the presence of seasonality. Similarly, Fourier or spectral (or frequency domain) methods also deserve more discussion. My own prejudice is that seasonality is usually obvious enough not to need discovery as a massive spike in the spectrum. Nevertheless, sometimes only spectral methods can give the full context of variability at a range of frequencies. Newton (1993) surveyed graphics for time series, discussing frequency domain displays in some detail.

3   The Bills of Mortality

Bills of Mortality were issued weekly in London from the 16th century on giving counts of deaths from various causes, collating data from the several parishes in the city. They stimulated John Graunt (1620–1674), a London draper, to write *Natural and Political Observations . . . upon the Bills of Mortality*, one of the founding documents of statistics, epidemiology, and demography. He was elected to the then-young Royal Society within weeks of the book's publication.

From the fifth (and posthumous) edition of 1676, we take data on deaths from plague in various years, noting the peaks around August and September. Figure 1 shows the annual series superimposed, and figure 2 shows them separated. Logarithmic scales seem especially appropriate for explosive phenomena such as plague.

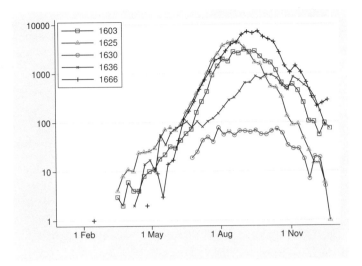

Figure 1: Plague deaths in London in various years from data reported by Graunt (1676). Note the shared tendency to peaks around August and September.

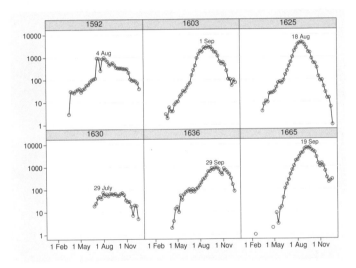

Figure 2: Plague deaths in London in various years from data reported by Graunt (1676). Added dates show weekly reports with highest numbers in each year.

In his edition of Graunt (1676), Hull gave detailed comments on the data. Implausible numerical quirks imply that the 1592 data are unreliable. Other sources indicate various small corrections and qualifications for the later years. However, none of these problems affect the main argument here.

Choosing between superimposing and juxtaposing is not always easy. Although examples clearly give complementary views of a given dataset, you may not be able to persuade reviewers or editors to include both in a publication.

4 Stata tips for plotting versus time of year

Reviewing some small but practical points for graphs of this kind may be helpful. The data may have arrived as, or been converted to, Stata date variables, but having, e.g., separate month and year variables is also helpful.

An especially useful function is `doy()` for day of year, running from 1 to 365 or 366. Note also the `egen` function `foy()` for fraction of year in the `egenmore` package on SSC (see [R] **ssc** for more on SSC).

Check out built-in sequences, such as `c(Mons)`. See the results of `creturn list`, scrolling toward the end. See also Cox (2004a).

Remember `twoway connected` as well as `line`. Although line plots are conventional in various disciplines, connected plots have the merit of showing individual data points. Marker symbol size can always be tuned to be noticeable but not obtrusive.

Use the `separate` command to separate one variable into several for easy comparison. See also Cox (2005b) for another example.

Because zeros cannot be shown as such on logarithmic scales, change zeros to missing in a copy of the data. Then prohibit connections across spells of missing values with the option `cmissing(n)`.

5 Cycle plots

5.1 Introduction

Graunt's data come for selected years. Having single or multiple time series extending over several years is more common. Figure 3 is an example from economics with monthly data. Trend, seasonality, and irregularities (attributable here mostly to strikes) are all evident. The data are for distance flown by U.K. airlines and come from Kendall and Ord (1990). Logarithmic scales again appear natural.

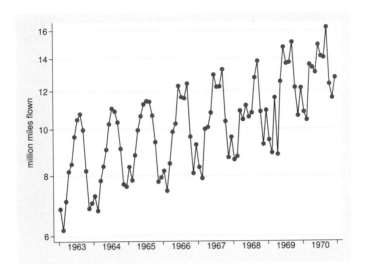

Figure 3: Distance flown by U.K. airlines—a common kind of economic time-series graph showing trend, seasonality, and irregularities.

This graph illustrates an elementary principle: the **sort** order for monthly data is naturally first by year and then by month. The idea of cycle plots is just to reverse that: sort by month, and then by year, to see the information in a different way. We could do this by using some **graph** command and an option, **by**(*monthvar*), but there would be too much scaffolding. Hence I have written **cycleplot** for this purpose and formally publish it with this column.

5.2 Syntax

cycleplot *responsevars month year* [*if*] [*in*] [, l̲ength(#) s̲tart(#)
   s̲ummary(*egen_function*) m̲ylabels(*labels_list*) *line_options*]

5.3 Options

l̲ength(#) indicates that data are for # shorter periods within each longer period. The default is 12, for months within a year.

s̲tart(#) indicates the first value of month plotted on the *x*-axis. The default is **start(1)**. This option may be used whenever there is some better natural start to the year than (say) January. For example, rainfall in climates with a wet season either side of December is best plotted starting in (say) July.

s̲ummary(*egen_function*) calculates a summary function to be shown for each month. The summary function may be any function acceptable to **egen** that has syntax like **egen** *newvar* = **mean**(*response*), **by**(*month*). **mean**() and **median**() are the most obvious possibilities. Know that whenever summaries are plotted, the order of variables on the graph is all the response variables followed by all the corresponding summary variables.

mylabels(*labels_list*) specifies text labels to use on the time axis, instead of default labels such as 1/12. The number of labels specified should be the same as the argument of length(), or by default 12. Labels consisting of two or more words should be bound in " ". Labels including " should be bound in ' " " '. mylabels('c(Mons)') specifies Jan Feb Mar ... Nov Dec, and mylabels('c(Months)') specifies January February March ... November December. Do not rotate the list to reflect a start() choice other than 1; this step will be done automatically.

line_options refers to options of graph twoway line; see [G-2] **graph twoway line**. connect(L ..) is wired in. You can use recast() to get a different twoway type.

5.4   Examples

Cycle plots have been discussed under other names in the literature, including cycle-subseries plot, month plot, seasonal-by-month plot, and seasonal subseries plot. For textbook treatments, see Becker, Chambers, and Wilks (1988); Cleveland (1993, 1994); or Robbins (2005). For research paper examples, see Cleveland and Devlin (1980); Cleveland and Terpenning (1982); Cleveland, Freeny, and Graedel (1983); or Cleveland et al. (1990).

Figure 4 is a default cycle plot for our example. We see the structure of seasonality much more easily, especially details such as the shift in peak from July to September.

The syntax used was

```
. cycleplot air month year,
> ylabel(6000 "6" 8000 "8" 10000 "10" 12000 "12" 14000 "14" 16000 "16", ang(h))
> ytitle(million miles flown) yscale(log)
```

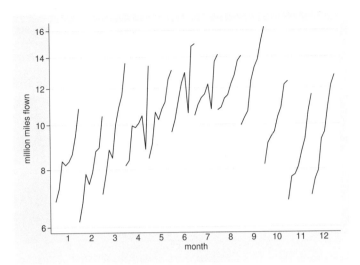

Figure 4: Distance flown by U.K. airlines. This cycle plot gives a different take on seasonality, more clearly showing timing (and shifts in timing) of peaks and troughs.

The program `cycleplot` can plot several responses and is applicable to any setup of longer periods divided into a fixed number of shorter periods. Quarterly data are thus another application. We will stick to the terms "month" and "year" as more concise, despite the imprecise terminology.

In `cycleplot`, you can rotate the time axis to start within the year. Experience indicates that splitting troughs, not peaks of the cycle, is best, although the opposite would apply if troughs were the focus of interest. Thus in studying rainfall variations, split the dry season rather than the wet season, unless the structure of the dry season is of concern.

You can also superimpose a summary for each month by naming the corresponding `egen` function, such as `mean`.

Standard graph options include `recast()`. Figure 5 shows the previous cycle plot, modified merely by adding the option `recast(connected)` and tweaking the axis labels by the option `mylabels('c(Mons)')`.

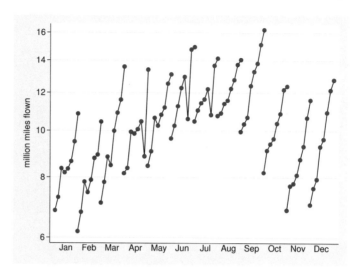

Figure 5: Distance flown by U.K. airlines. This cycle plot has been tweaked into a connected plot, and the month axis labels have been modified.

Here is another example, from medical statistics. Figure 6, using data from Diggle (1990), shows deaths in the United Kingdom from bronchitis, emphysema, and asthma. Seasonality is no surprise here, but as before a cycle plot is better than the standard time-series plot at showing the fine structure—indeed at showing basic details such as peak and trough months. A logarithmic scale makes each fluctuation up or down come out around the same height. Figure 7 shows a cycle plot, here rotated so that the winter is not cut, by using the option `start(8)`, and recast as a connected plot, by using the option `recast(connected)`.

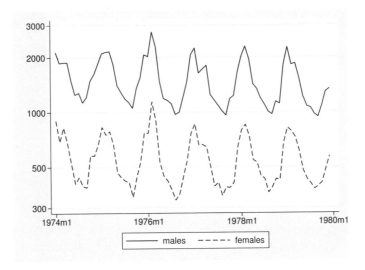

Figure 6: Deaths in the United Kingdom from bronchitis, emphysema, and asthma. Standard line plot of a strongly seasonal series.

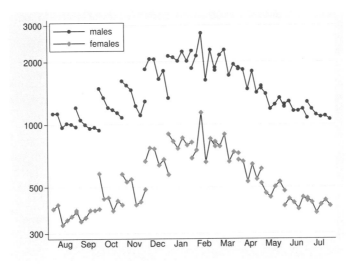

Figure 7: Deaths in the United Kingdom from bronchitis, emphysema, and asthma. This cycle plot more clearly shows the structure of seasonality.

6   Do-it-yourself rotation

cycleplot allows you to rotate the time-of-year axis. Few analysts will need much convincing that rotation can be a good idea. So how could you do it yourself?

Let us keep the example of monthly data and assume that a `month` variable runs from 1 (January) to 12 (December). (Separate month and year variables are useful even when you have Stata date variables.) Say that you want to start the year in month 8 (August). So months 8–12 are to be mapped to positions 1–5, and months 1–7 are to be mapped to positions 6–12.

An expression to use in generating such a new variable is

```
cond(month > 7, month - 7, month + 5)
```

as there are two cases to cover, the second part of the year that becomes the first and vice versa. See Kantor and Cox (2005) for a tutorial on `cond()`. An alternative is

```
1 + mod(month - 8,12)
```

as the remainder on dividing integers by 12 must vary from 0 to 11. I suggest that the latter method is more elegant but the former is easier to emulate.

Short of fixing axis labels, that is all that you need to know. However, you might wish to note various pertinent `egen` functions in Cox (1999, 2000) and `egenmore` from SSC.

7 Mauna Loa: Superimposing, slicing, stacking

7.1 Introduction

In 1958 the oceanographer Charles D. Keeling (1928–2005) started what is now the longest continuous series of carbon dioxide measurements on top of Mauna Loa, Hawaii. This dataset is crucial to discussions of human effects on the atmosphere. The units are ppm, parts per million (by volume). Thus 300 ppm = 0.03%.

I accessed data from http://cdiac.ornl.gov/ftp/trends/co2/maunaloa.co2 on March 22, 2006 and linearly interpolated a few small gaps in the early part of the record. Figure 8a shows a strong trend and seasonality. Given the trend, a plot against month using `connect(L)` is interesting (figure 8b). The lack of overlap here can be considered fortuitous but also fortunate. `connect(L)` connects if and only if the x-axis variable is increasing (strictly, not decreasing). `connect(l)` would be useless here, producing logical but confusing backward connections between each December (12) and the following January (1).

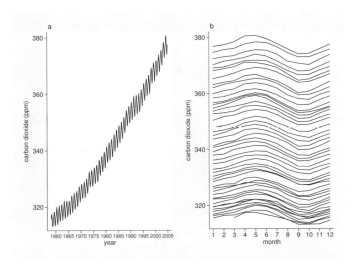

Figure 8: (a) Carbon dioxide measured at Mauna Loa shows a strong upward trend and fairly systematic seasonality. (b) Plotting against time of year gives a handle on the seasonality. By chance, no playing with offsets is needed for the annual segments.

Given such series, we should smooth or model and look at the residuals. How best to do that is a fascinating subject, and time-series experts could have a field day comparing their favorite methods, but here we just use the `lowess` default and plot the residuals from that. A superimposed line plot (figure 9a) and a standard time-series plot (figure 9b) of residuals show the family resemblance of seasonal cycles, but whether you choose spaghetti or a roller-coaster, each shows a clear pattern but also fails to suggest anything new.

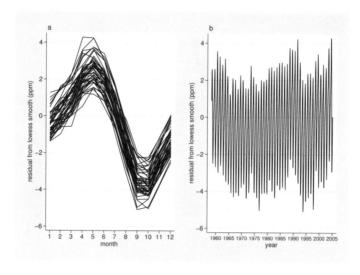

Figure 9: (a) Residuals from `lowess` default plotted against time of year. (b) Same residuals plotted as time series.

In particular, the aspect ratio of figure 9b is a problem. Standard advice (Fisher 1925; Cox 2004b) is to choose an aspect ratio such that line segments are as near 45° as possible, but here that would lead to a long graph. An alternative is to slice the series into parts, graph each part, and then stack the graphs by using `graph combine`. The details are mostly mundane but typically tedious. `sliceplot`, here published formally, is a wrapper program to automate that process.

7.2 Syntax

`sliceplot` *plottype yvarlist xvar* [*if*] [*in*] [, `at`(*numlist*) `unequal` `length`(*#*)
    `slices`(*#*) `combine`(*combine_options*) *twoway_options*]

7.3 Options

`at`(*numlist*) specifies cutpoints for the ends of each slice as values of the x-axis variable. Values outside the range of the data will be ignored with a warning.

`unequal` may be used with `at()` if you want to allow slices to have unequal scales. It specifies that unequal scales be used on slices of different length. The default is to use (approximately) the same scale. A common application is to show more interesting values at a greater magnification than others.

`length`(*#*) specifies the maximum length of each slice in units of the x-axis variable. The default is `length(100)`.

slices(*#*) specifies the number of slices.

combine(*combine_options*) specifies options of graph combine; see [G-2] **graph combine**. The defaults are imargin(zero) cols(1).

twoway_options are options of graph twoway (see [G-2] **graph twoway**) controlling other features of the graph.

7.4 Examples

Figure 10 shows an example of what sliceplot can do.

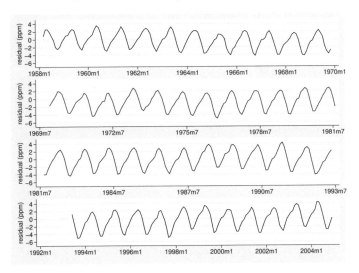

Figure 10: Residuals from lowess default plotted in slices to give a more congenial aspect ratio.

The command for that is

```
. sliceplot line res date, slices(4) ytitle(residual (ppm))
> ylabel(-6(2)4, angle(h)) xtitle("")
```

showing that sliceplot is a wrapper command that calls up a graphics command and slices the dataset by cutting the horizontal axis. You may specify both slicing options and standard graph options. Here we ask for just four slices, but options also exist to control slice endpoints and lengths. An analog could be written to cut the vertical axis, but I find that this aspect ratio problem occurs mostly with time series.

8 Loops in state space

One basic technique—perhaps more common in physics than in mainstream statistics—is to consider plots in some state space. Figure 11a is a basic line plot of residual versus previous residual for the Mauna Loa data. `lwidth(0)` (indeed) is a way to get thin lines. Figure 11b shows that we can identify months, which underlines the regularity of this cycle.

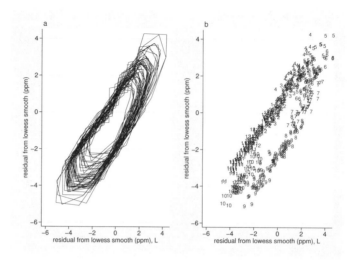

Figure 11: Residuals versus previous residuals shown using (a) a connected line and (b) month identifiers.

We can also connect with arrows by using `twoway pcarrow`. The main idea here was discussed in detail in Cox (2005a). Figure 12 gives another handle showing more of the repetitive fine structure of each seasonal cycle.

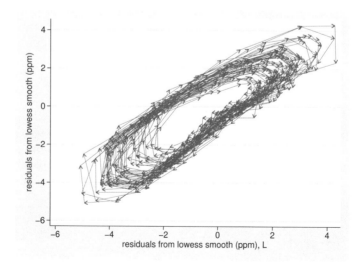

Figure 12: Residuals versus previous residuals shown using arrows.

For another application of the state space idea, let us revisit one of the staples of elementary geography, graphs of monthly means of precipitation and temperature. The usual graphs cut the year, sometimes painfully. Figures 13 and 14 give conventional graphs of the seasonal cycle for Boston, Houston, and San Francisco in the United States, using data from Pearce and Smith (1984). In the dataset, these cities are separate panels.

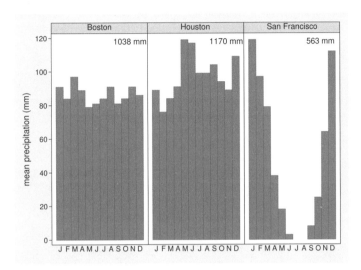

Figure 13: Annual cycle of precipitation for Boston, Houston, and San Francisco. Annual totals shown by text.

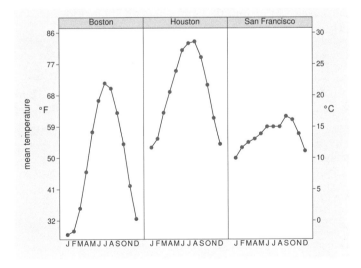

Figure 14: Annual cycle of temperature for Boston, Houston, and San Francisco.

One of various alternatives to the usual graphs is to plot the annual cycle as a loop in some two-dimensional space, say, combining precipitation and temperature. Such graphs are often called climagraphs or climographs, but there is nothing intrinsically climatic about them. It appears (Linacre 1992) that they go back to Alexandre Gustave Eiffel (1832–1923), better known for more towering achievements. For examples in a medical context, see Cliff, Haggett, and Smallman-Raynor (2004).

Figure 15a is an example in which the monthly means from January to December are connected in time order. However, December logically should also be connected to January to close the loop. Figure 15b is the result.

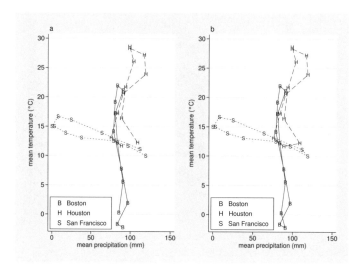

Figure 15: Annual cycle of precipitation and temperature for Boston, Houston, and San Francisco. (a) Open loop. (b) Closed loop.

How did we do that? We need to add an extra observation at the end of each panel that is a copy of the first observation. The main idea is to use `by:` and `expand`.

In more detail: The structure of the dataset is three panels and 12 months for each panel. We need to tag the first observation in each panel and then create a copy of those first observations. Knowing that `expand` adds extra observations at the end of the dataset helps. Each extra observation is assigned a value of `month` of 13, which ensures that after `sorting`, the new observation will be in the right position.

```
. preserve
. local N = _N
. by place (month), sort: generate first = _n == 1
. expand 2 if first
. replace month = 13 if _n > `N´
. sort place month
. graph_commands
. restore
```

Here we `preserve` and then `restore` so that the original dataset is in memory after graphics. Other solutions to the problem caused by a modification of the data, which we want only for this purpose, include a `save` of the original dataset so that it can be returned to as and when desired.

9 Incidence plots

What are here called incidence plots are scatterplots of the form

```
scatter year month if
```
condition

`year` and `month` are named here for concreteness. Your names naturally may differ, and your month variable may even be day of year, quarter, or some other suitable time unit. Whichever variables you choose, such an incidence plot is in essence a graphical table in which each year is a row. Logically equivalent is a scatterplot of the form

```
scatter month year if
```
condition

in which each year is a column.

As we can superimpose several such plots, we can compare different years, even in a fairly long time series, with a bird's-eye view of the incidence of several different *condition*s.

The Mauna Loa data have been `tsset`, so we can use time-series operators, for example to look at changes from value to value. So after

```
. summarize D.co2, detail
```

we can show months with large positive changes (say, those in the top 10%) and months with large negative changes (say, those in the bottom 10%). The result is given in figure 16.

```
. scatter year month if D.co2 > `=r(p90)´, options
> ||
> scatter year month if D.co2 < `=r(p10)´, more_options
```

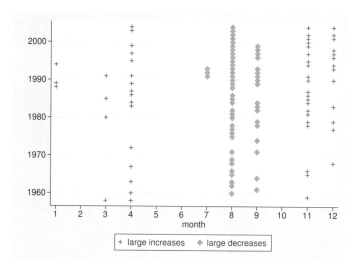

Figure 16: Incidence plot showing months of largest increases and decreases in carbon dioxide content at Mauna Loa.

Sakamoto-Momiyama (1977) makes good use of a related idea. Her disease calendars use a series of bar charts to show months of highest mortality for various diseases for different years, age groups, countries, etc. This information is within a monograph that is dense with a variety of carefully designed graphics to show seasonal variations in mortality.

10   Folding

The time-of-year axis can be folded so that the second half of the year is superimposed on the first, giving more space and a graphical handle on the asymmetry of annual cycles.

With monthly data, folding is best accomplished by the transformation `min(month, 14 - month)`, which pairs months as follows: 1 by itself, 2 and 12, 3 and 11, 4 and 10, 5 and 9, 6 and 8, and 7 by itself. Naturally a similar transformation may be used after a rotation.

Folding in this manner was used by the climatologist Victor Conrad (1876–1962). See Conrad and Pollak (1950).

11   Repeating

Values in the latter part of the year can be copied left of the start, and values in the earlier part of the year can be copied right of the end. This method reduces the effects of cutting. Mathematician and scientist Johann Heinrich Lambert (1728–1777) used repeating in this manner with seasonal data. Tufte (2001, 29) accessibly reproduces an example graph. More recently, Tukey (1972) blew a trumpet for the idea that two cycles are better than one. Two cycles are naturally not compulsory: you can copy as much or as little as desired.

The Stata code for this process is a variation on that given earlier for adding extra observations to close loops by connecting the last and first in each panel. It can be done using `expand`, often after `preserve` and before `restore`. One sequence could run like this, for two cycles:

```
. preserve
. local N = _N
. expand 2 if month <= 6
. replace month = month + 12 if _n > `N'
. local N = _N
. expand 2 if month >= 7
. replace month = month - 12 if _n > `N'
. graph_commands
. restore
```

This code gives two cycles of monthly data. First, the first 6 months are copied, and in the copies, months 1–6 are mapped to 13 to 18. Then the last 6 months are copied, and in the copies, months 7–12 are mapped to −5 to 0. The correct sort order for the graph can be obtained by an explicit `sort` or on the fly by a `sort` option of `graph`. Panel data need use of `by:`, as seen earlier.

Figure 17 reunites San Francisco's wet winter.

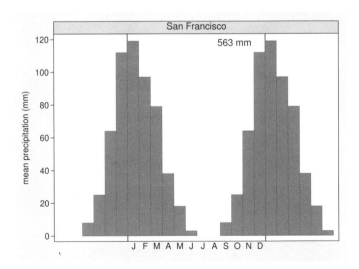

Figure 17: Annual cycle of precipitation in San Francisco. Each month is shown twice. Annual total shown by text.

12   Conclusion

For seasonal data, I give this advice on graphics.

Graphs showing the fine structure of seasonality tell us more than graphs that serve mostly to reveal its existence. The examples here are of well-understood phenomena. Can you use the method to break new ground in understanding fresh datasets?

Reordering the data into subseries (`cycleplot`) is often useful; rotate to start at an appropriate time of year for the analysis; superimpose, slice, and stack to compare years (`sliceplot`); plot loops in state space; use incidence plots; fold the time-of-year axis; and repeat values fore and aft to show up to two cycles.

Know your functions, graphics options, and data management commands. Each new program can be a curse as well as a convenience, being just one more thing to learn, remember, forget, and confuse. Once you understand the logic for rotating axes or repeating values fore and aft, the need for extra commands or extra functions to do such tasks diminishes rapidly.

13   Acknowledgments

Aurelio Tobias contributed to the development of `cycleplot`. Marcello Pagano, Austin Nichols, and Joe Newton supplied valuable references.

14   References

Becker, R. A., J. M. Chambers, and A. R. Wilks. 1988. *The New S Language: A Programming Environment for Data Analysis and Graphics*. Pacific Grove, CA: Wadsworth and Brooks/Cole.

Brinton, W. C. 1914. *Graphic Methods for Presenting Facts*. New York: Engineering Magazine Company.

Cleveland, R. B., W. S. Cleveland, J. E. McRae, and I. Terpenning. 1990. STL: A seasonal-trend decomposition procedure based on loess. *Journal of Official Statistics* 6: 3–73.

Cleveland, W. S. 1993. *Visualizing Data*. Summit, NJ: Hobart.

———. 1994. *The Elements of Graphing Data*. Rev. ed. Summit, NJ: Hobart.

Cleveland, W. S., and S. J. Devlin. 1980. Calendar effects in monthly time series: Detection by spectrum analysis and graphical methods. *Journal of the American Statistical Association* 75: 487–496.

Cleveland, W. S., A. E. Freeny, and T. E. Graedel. 1983. The seasonal component of atmospheric CO_2: Information from new approaches to the decomposition of seasonal time series. *Journal of Geophysical Research* 88: 10934–10946.

Cleveland, W. S., and I. J. Terpenning. 1982. Graphical methods for seasonal adjustment. *Journal of the American Statistical Association* 77: 52–62.

Cliff, A. D., P. Haggett, and M. Smallman-Raynor. 2004. *World Atlas of Epidemic Diseases*. London: Arnold.

Conrad, V., and L. W. Pollak. 1950. *Methods in Climatology*. Cambridge, MA: Harvard University Press.

Cox, N. J. 1999. dm70: Extensions to generate, extended. *Stata Technical Bulletin* 50: 9–17. Reprinted in *Stata Technical Bulletin Reprints*, vol. 9, pp. 34–45. College Station, TX: Stata Press.

———. 2000. dm70.1: Extensions to generate, extended: corrections. *Stata Technical Bulletin* 57: 2. Reprinted in *Stata Technical Bulletin Reprints*, vol. 10, p. 9. College Station, TX: Stata Press.

———. 2004a. Stata tip 9: Following special sequences. *Stata Journal* 4: 223.

———. 2004b. Stata tip 12: Tuning the plot region aspect ratio. *Stata Journal* 4: 357–358.

———. 2005a. Stata tip 21: The arrows of outrageous fortune. *Stata Journal* 5: 282–284.

———. 2005b. Stata tip 27: Classifying data points on scatter plots. *Stata Journal* 5: 604–606.

———. 2006. Speaking Stata: Time of day. *Stata Journal* 6: 124–137.

Diggle, P. J. 1990. *Time Series: A Biostatistical Introduction*. Oxford: Oxford University Press.

Fisher, R. A. 1925. *Statistical Methods for Research Workers*. Edinburgh: Oliver & Boyd.

Graunt, J. 1676. *Natural and political observations, mentioned in a following index, and made upon the bills of mortality*. London: John Martyn. Reprinted in *The Economic Writings of Sir William Petty, Together with the Observations upon the Bills of Mortality, More Probably by Captain John Graunt*, 1899, ed. C. H. Hull. Cambridge: Cambridge University Press.

Kantor, D., and N. J. Cox. 2005. Depending on conditions: A tutorial on the cond() function. *Stata Journal* 5: 413–420.

Kendall, M. G., and J. K. Ord. 1990. *Time Series*. London: Arnold.

Linacre, E. T. 1992. *Climate Data and Resources: A Reference and Guide*. London: Routledge.

Newton, H. J. 1993. Graphics for time series analysis. In *Handbook of Statistics 9: Computational Statistics*, ed. C. R. Rao, 803–823. Amsterdam: North-Holland.

Pearce, E. A., and C. G. Smith. 1984. *The World Weather Guide*. London: Hutchinson.

Robbins, N. M. 2005. *Creating More Effective Graphs*. Hoboken, NJ: Wiley.

Sakamoto-Momiyama, M. 1977. *Seasonality in Human Mortality: A Medico-geographical Study*. Tokyo: University of Tokyo Press.

Tufte, E. R. 2001. *The Visual Display of Quantitative Information*. 2nd ed. Cheshire, CT: Graphics Press.

Tukey, J. W. 1972. Some graphic and semigraphic displays. In *Statistical Papers in Honor of George W. Snedecor*, ed. T. A. Bancroft and S. A. Brown, 293–316. Ames, IA: Iowa State University Press.

About the author

Nicholas Cox is a statistically minded geographer at Durham University. He contributes talks, postings, FAQs, and programs to the Stata user community. He has also coauthored 15 commands in official Stata. He was an author of several inserts in the *Stata Technical Bulletin* and is an editor of the *Stata Journal*.

The Stata Journal (2007)
7, Number 3, pp. 413–433

Speaking Stata: Turning over a new leaf

Nicholas J. Cox
Department of Geography
Durham University
Durham City, UK
n.j.cox@durham.ac.uk

Abstract. Stem-and-leaf displays have been widely taught since John W. Tukey publicized them energetically in the 1970s. They remain useful for many distributions of small or modest size, especially for showing fine structure such as digit preference. Stata's implementation `stem` produces typed text displays and has some inevitable limitations, especially for comparison of two or more displays. One can re-create stem-and-leaf displays with a few basic Stata commands as scatterplots of stem variable versus position on line with leaves shown as marker labels. Comparison of displays then becomes easy and natural using `scatter, by()`. Back-to-back presentation of paired displays is also possible. I discuss variants on standard stem-and-leaf displays in which each distinct value is a stem, each distinct value is its own leaf, or axes are swapped. The problem shows how one can, with a few lines of Stata, often produce standard graph forms from first principles, allowing in turn new variants. I also present a new program, `stemplot`, as a convenience tool.

Keywords: gr0028, stemplot, stem-and-leaf, graphics, distributions, digit preference

1   Introduction

I was attracted to stem-and-leaf displays when I first learned about them in the mid-1970s from John Tukey's writings. They were a fixture in my introductory courses from the start of my teaching career. But over the last decade or so, I have come to use them less. As usually implemented, they are less flexible than dot or strip displays. This column is a story of how one can produce stem-and-leaf displays by using Stata's graphics, thus increasing their versatility. Within this story, other questions will emerge: How far can you get using a few basic commands? When is it better to wrap up those commands within a program?

2   Stem-and-leaf displays

Just in case you have missed out on stem-and-leaf displays, let us run through the basic idea and its implementation in Stata as `stem`.

The name *stem-and-leaf* is attributable to Tukey (1972, 1977). Their widespread use is undoubtedly the result of his energetic and enthusiastic advocacy. The basic idea appears in other places, including Dudley (1946, 22), as reproduced by Emerson and

Hoaglin (1983, 19), and Japanese railway timetables, as reproduced by Tufte (1990, 45–46). Among many textbook accounts are those of Griffiths, Stirling, and Weldon (1998); Wild and Seber (2000); and Moore and McCabe (2006).

Let us look at some examples.

```
. sysuse auto
(1978 Automobile Data)
. stem length

Stem-and-leaf plot for length (Length (in.))
  14* | 279
  15* | 45567
  16* | 133455589
  17* | 00002234457999
  18* | 02469
  19* | 23356788889
  20* | 00001113446667
  21* | 224788
  22* | 00112
  23* | 03
```

Each value (e.g., 142) is split into so-called *stem* (e.g., 14) and *leaf* (e.g., 2) parts. Here each distinct stem is used just once, but there is no rule about that. Another example shows how stems are repeated on two or more lines:

```
. stem mpg

Stem-and-leaf plot for mpg (Mileage (mpg))
  1t | 22
  1f | 44444455
  1s | 66667777
  1. | 88888888899999999
  2* | 00011111
  2t | 22222333
  2f | 444455555
  2s | 666
  2. | 8889
  3* | 001
  3t |
  3f | 455
  3s |
  3. |
  4* | 1
```

Here, at least for English speakers, accidents of language make t, f, and s easy labels for two or three, four or five, and six or seven, respectively. In general, a stem may be used most easily one, two, or five times with decimal representations.

2.1   A note on units

In most countries in the world, the units used in examples in this column are not standard, so I will explain. The car lengths in the previous example are measured in inches. An inch is defined to equal 25.4 mm. Twelve inches = 1 foot, 5,280 feet = 1 mile, and 1 U.S. gallon is about 3.785 liters. Hence 1 mile per gallon (mpg) is about 0.425

km per liter or 2.352 liter per km. Twenty miles per gallon, the median in the `auto` dataset, is about 8.50 km per liter. Pennycuick (1988) and Darton and Clark (1994) are two of many useful references on units of measurement.

2.2 Advantages

The display is a graphical condensation of the list of ordered values 142, 147, 149, ..., 230, 233. Once you realize that, its key features are immediate, which is much of the attraction:

- *A low-technology method.* You can do stem-and-leaf displays by hand, using any convenient piece of paper. (Paper with large squares keeps things tidy.) Evidently, that consideration weighs less and less in an era when many readers take laptops with them wherever they go. But it still has weight, especially as a way of getting children and other learners to look at distributions in small, simple datasets. Similarly, one can easily type or print a stem-and-leaf display. That was an advantage to many in the mid-1970s, but again it counts for little 30 years later. Indeed, handling typed displays in some fixed-width font may now be less familiar and more awkward for you than handling graphic images.

- *Keeps most or all information.* Each value is split into stem and leaf parts. For `length`, there is no more detail, so the stem-and-leaf display with leaves that are single digits loses no information about magnitudes. Whenever that is not true, there are various possibilities, including rounding the data and using leaves that are two or more digits.

- *Shows distributions in histogram style, but including fine structure.* The ordering and grouping yield a histogram on its side, which is, presumably, why stem-and-leafs are usually shown with magnitude increasing down the page. A rotation of 90° counterclockwise yields a histogram. Demonstrating this effect physically by rotating an acetate transparency in the classroom is, unfortunately, now regarded as technologically incorrect. The stem-and-leaf display, however, retains more detail (as seen, often all the detail) of individual data points. That quality is often useful, especially for inspecting outliers and other extremes and any granularity in the distribution, such as may arise from digit preference (Preece 1981). A display need not appear in published reports for it to be useful in exploratory checking of your data.

- *Allows calculation of measures on the basis of ordered values.* The ordering allows fairly easy hand calculation of summary measures on the basis of order statistics (median, quartiles, and so forth). Tukey's low-technology definitions of what later were called letter values (Hoaglin 1983) involved nothing more complicated than splitting the difference between adjacent order statistics. (Doing so makes the definitions trickier to extend to variables associated with weights.) This detail still deserves brief emphasis in introductory texts and courses but is of little importance for researchers able to calculate such measures directly with software such as Stata.

2.3   Limitations

The stem ([R] **stem**) command was first introduced in Stata 3.0 in 1992. It does what it does well and has smart defaults based on the distribution of each variable. Its main features include handles for tuning the display and the possibility of repeating displays for distinct groups with a by() option.

Given these advantages, why are stem-and-leaf displays not used more? The limitations are just about as obvious:

- *Problems with large datasets.* Stem-and-leaf displays work well for datasets of small or modest size but otherwise may easily be awkward, if not impracticable. The length of the longest line, the number of lines, and the legibility of leaves can quickly become problematic, and there are few easy trade-offs. Using some kind of dot plot (in the sense of [R] **dotplot**, not [G-2] **graph dot**) is then more practical: we must give up on the idea of a leaf and use some small point symbol instead. Alternatively, many people would turn to histograms, kernel density estimates, quantile plots, or other representations.

- *Are final digits interesting?* The last digit(s) may often not be at all interesting or informative, especially when data were produced by some automated measurement procedure. (Whenever people produce numbers, there is a chance of fine structure, perhaps trivial, as a result of digit preferences or other quirks.) If the detail of leaves is noise to the user, then the simplicity of dot plots is again preferable.

- *Comparison can be awkward.* Comparison of two or more stem-and-leaf displays can be difficult or inefficient unless they are displayed side by side with a common magnitude scale. The action of stem, almost inevitably, is to emit displays in sequence, so only reediting in text or word processing software could achieve this. Aligning displays, line length, and legibility of leaves could again prove problematic.

3   Stem-and-leaf displays as scatterplots

These limitations are serious, but they do not rule out the usefulness of stem-and-leaf displays in all circumstances. Moreover, some are largely consequences of using printed text to show stem-and-leaf displays as, in essence, tables rather than graphs. For example, keeping track of various small tables as bundles of characters and trying to emit them properly aligned would be a small nightmare, but getting small graphs aligned as you wish is easy, because you just tell Stata to do it.

A fresh look at the problem grows from the realization that stem-and-leaf displays are essentially a kind of scatterplot. The ingredients include the following:

1. The variable in question, or a rounded variant of it, gives the y coordinate. For that, we may need to learn a little about how to round.

2. The position of each leaf on each line gives the x coordinate. We will have to construct this. A natural start is just to ensure that the first, second, and later leaves on each line are plotted against 1, 2, and so forth. (Only for one purpose will we need to do anything more complicated.)

3. A leaf must be shown at each point (x, y), rather than a point symbol. For that, we may need to learn a little about how to extract the needed digits. Leaves will be shown as marker labels, and marker symbols will be suppressed. If you are not familiar with marker labels, check out [G-3] *marker_label_options* or the equivalent online help.

In following this basic idea, we need not commit ourselves to producing exact replicas of typed or printed displays. Using graphics rather than text may well mean that different design choices are easier, more attractive, or more informative.

3.1 From first principles

The official `dotplot` ([R] **dotplot**) command yields displays similar to stem-and-leaf displays, but showing point symbols rather than leaves. `stripplot` on SSC is a loosely similar user-written command. So, we could try tricking either into showing leaves. A more direct approach is to try building up a command for ourselves from first principles. This strategy strengthens understanding and often shows that a few simple commands will give us the result we desire, thus avoiding the writing of a program.

My habit is to start with small toy examples and get something simple working quickly, not to play with the dataset I really want to work with (which is likely to be much larger), or to sketch an elaborate design in advance. (This is not necessarily good advice for large, complicated programming projects.) In that spirit, consider `mpg` in the `auto` dataset, the second stem-and-leaf example above. The values are all integers, ranging from 12 to 41. The simplest choice, at first sight, is to use the second digits as leaves. One way to do that is to extract the second digit as the remainder on division by 10:

```
. generate leaf = mod(mpg, 10)
```

Another way is to extract it as the last character of the string equivalent of `mpg`:

```
. generate leaf = substr(string(mpg), -1, 1)
```

Here `-1` as the second argument indicates the last character in a string. Naturally,

```
. generate leaf = substr(string(mpg), 2, 1)
```

would produce the same result with `mpg`, but one needs only a little foresight to see that this is less general, applying only to two-digit integers. `scatter` will happily use either numeric or string variables as marker labels, so the choice is for us to make on other grounds.

We need x and y coordinates. We need an x coordinate more, so let's get that first, and just use each distinct value as its own stem:

```
. by mpg, sort: generate x = _n
```

Under by:, the observation number _n is evaluated separately within distinct groups defined by its argument. Within groups of distinct values of mpg, it thus takes on the values 1, 2, and so forth, exactly as desired. If you want a more detailed tutorial on by:, see Cox (2002).

Figure 1 shows our initial stem-and-leaf plot:

```
. scatter mpg x, mla(leaf) mlabpos(0) ms(none)
```

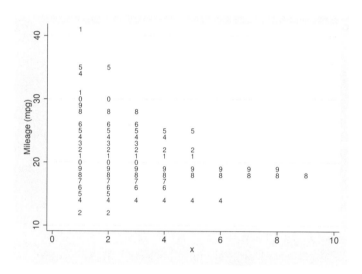

Figure 1: Stem-and-leaf plot for miles per gallon with scatter. Each distinct value serves as a stem. Final digits serve as leaves.

When we suppress marker symbols and show marker labels instead, using mlabpos(0) to center labels on where symbols would have been is often best.

The result is not especially pretty. More importantly, we are most of the way to where we want to be after just three lines. My first thoughts on this graph are in terms of simple cosmetic improvements. I do not want the x-axis title. I do not want the x-axis labels (although others might want a frequency scale on that axis). I do want horizontal y-axis labels. To match stem-and-leaf conventions, I do not want them to be ticked. I do not want the associated grid lines (although this preference also is a matter of taste). With a few minor changes, we produce figure 2.

```
. scatter mpg x, mla(leaf) mlabpos(0) ms(none) yla(, ang(h) noticks nogrid)
> xla(none) xti("")
```

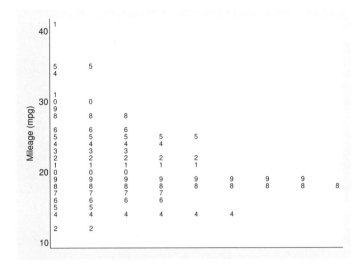

Figure 2: Same as figure 1 but with cosmetic improvements to axis labels and titles

If you are working through this sequence with your copy of Stata, you might be using a graph scheme that sets up a default of yla(, ang(h)), but making that choice explicit will do no harm.

My next thoughts are that I have more space to play with than I expected. Figure 3 suggests a variant in which each value is its own leaf:

```
. scatter mpg x, mla(mpg) mlabpos(0) ms(none) yla(, ang(h) noticks nogrid)
xla(none) xti("")
```

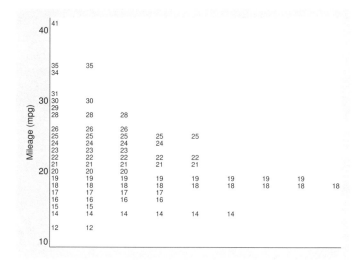

Figure 3: Same as figure 2, but now each value serves as its own leaf.

Now the *y*-axis labels are redundant, and we can take them off. If we do that, we will probably want to keep a gap in the same space, which is shown in figure 4.

```
. scatter mpg x, mla(mpg) mlabpos(0) ms(none) xla(none) xti("") yla(none)
> ysc(titlegap(4))
```

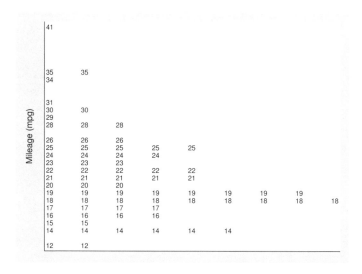

Figure 4: Same as figure 3, but now *y*-axis labels have been removed.

Our result is no longer a conventional stem-and-leaf display, but no matter: it is showing the same information in readable form.

Using distinct values as stems will work well only if there are enough ties, that is, not too many distinct values. (Some people say *unique values*, but for others that term means those values that occur just once.) You can see the number of distinct values by following a `tabulate` with `display r(r)`, because the number of rows in the table is saved immediately after a `tabulate`. `mpg` has 21 distinct values.

If you try this calculation with the first example above, `length`, the resulting graph (not shown here) is too messy. However, we can reuse most of the code. The main change is the need to recalculate the x coordinate.

```
. by length, sort: replace x = _n
. scatter length x, mla(length) mlabpos(0) ms(none) xla(none) xti("")
> yla(none) ysc(titlegap(4))
```

There are 47 distinct values of `length`. In general, performing `tabulate` beforehand on a potential stem variable will give a foretaste of the number of stems needed and their lengths. My own rule is that more than about 30 stems is usually too messy, unless you are happy to produce a tall graph and can convince those in power over you (advisors, supervisors, reviewers, editors) that it is a good idea.

3.2 Rounded stems

We now need to focus on how to produce rounded stems. `stem`'s choices for `length`, which varies between 142 and 233 inches, are 14(0), ..., 23(0), which are simple and appear good. Your first thought may be to extract the stem as the first two characters of the string equivalent of `length`, as, say, `substr(string(length), 1, 2)`, but this choice would pose two problems.

First, we need a numeric variable for graphing. That problem is easily soluble: convert the string to numeric by wrapping the expression with `real()` to give `real(substr(string(length), 1, 2))`.

Second, a deeper problem is that extracting the first few digits will give some wrong answers for any variable with different numbers of digits. (Consider the results if values varied between 142 and 2,330 instead.) Approaching the problem directly as a numeric calculation is better in general. Thus the stems for `length` are obtainable by dividing by 10 and rounding down:

```
. generate stem = int(length/10)
```

The `floor()` function would do as well as the `int()` function so long as values are not negative. However, if values are ever negative, `floor()`, which always rounds down, gives the wrong answer. `int()`, which always rounds towards zero, is preferable. The stem for -142 should be -14, not -15.

This rule would work well enough to re-create `stem`'s rendering of `length`, but we need a more general solution for whenever the width (in `stem`'s terminology) is not 10, or even any other exact power of 10. Suppose that we wanted twice as many stems than is given by 10. A rule that was

```
. generate stem = int(length/5)
```

would yield a stem of 28 for values 140–144, 29 for values 145–149, and so forth. But we would not want those stems to be shown as axis labels. A better command for graph purposes would be

```
. generate stem = width * int(varname / width)
```

So, we should start with

```
. generate stem = 10 * int(length/10)
```

However, there is a better approach that will save us work in the long run. Using `clonevar` ([D] **clonevar**) first,

```
. clonevar stem = length
. replace stem = 10 * int(length/10)
```

has a pleasant result. The effect of `clonevar` is that `stem` is born an exact copy of `length`, notably with the same variable label. That is a useful detail for graphing. Naturally, following `clonevar` with `replace` does not affect the variable label.

Let us put together in one place the code for a better stem-and-leaf plot for `length`, starting from nothing. We will need to calculate the x coordinate by using the variable `stem`—ensuring that the leaves are in the right order—and the leaf by using `length` itself. The code will produce figure 5.

```
. sysuse auto, clear
. clonevar stem = length
. replace stem = 10 * int(length/10)
. sort stem length
. by stem: generate x = _n
. generate leaf = substr(string(length), -1, 1)
. scatter stem x, mla(leaf) mlabpos(0) ms(none) yla(, ang(h) noticks nogrid)
> xla(none) xti("")
```

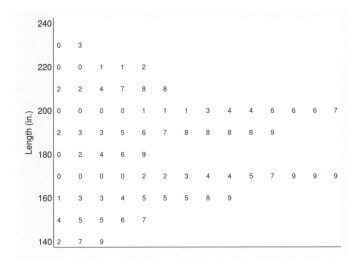

Figure 5: Stem-and-leaf plot for car lengths. Stems are based on intervals of width 10 in. Leaves are final digits.

From here, there are various ways to proceed. One is to make explicit all stems (and follow the convention of showing them truncated) by adding

```
yla(140 "14" 150 "15" 160 "16" 170 "17" 180 "18" 190 "19" 200 "20" 210
"21" 220 "22" 230 "23")
```

Another is to note that there is plenty of white space to spare. So we could use the values as their own leaves and suppress the y-axis labels:

```
. scatter stem x, mla(length) mlabpos(0) ms(none) yla(none) xla(none) xti("")
> ysc(titlegap(4))
```

3.3 Comparisons

One of the limitations of the **stem** command is that with a **by()** option, stem-and-leaf displays for different groups can be produced only in vertical sequence. Being able to compare displays that are aligned horizontally would be better, suggesting the use of **scatter** with a **by()** option. The only questions are what difference this might make to preprocessing and what extra details we might need to think about graphically.

Let us return to **mpg** and examine its variation with groups of **foreign** or **rep78**. For simplicity, values will be their own leaves. For clarity, we will start again from the top to show the full sequence of commands, which will produce figures 6 and 7.

```
. sysuse auto, clear
. by foreign mpg, sort: generate x = _n
. scatter mpg x, by(foreign, compact) xla(none) xti("") ms(none) mla(mpg)
> mlabpos(0) yla(none)
```

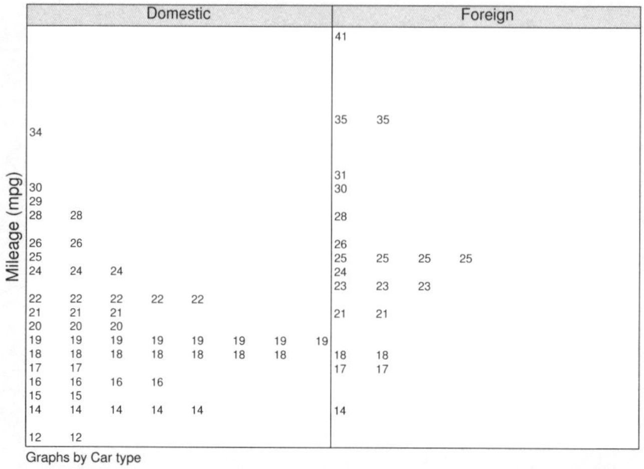

Figure 6: Stem-and-leaf plots of miles per gallon for domestic and foreign cars. Each value serves as its own leaf.

```
. by rep78 mpg, sort: generate X = _n
. scatter mpg X, by(rep78, compact row(1)) xla(none) xti("") ms(none)
> mla(mpg) mlabpos(0) yla(none)
```

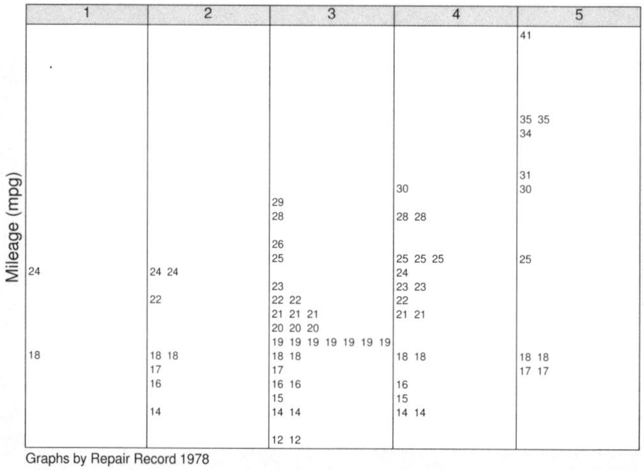

Figure 7: Stem-and-leaf plots of miles per gallon by repair record. Each value serves as its own leaf.

The calculation of the x coordinate needs to be done separately by the grouping variable. Suboptions of by() may come into play. compact and row(1) are useful. However, beware that total does not do what you might hope, because it necessarily superimposes different groups.

3.4 Back to back

When two stem-and-leaf plots are compared, some people like to plot them back to back. The result loosely resembles the pyramidal displays often used to compare age structure of populations for males and females. We could discuss whether the result is more attractive or more informative, but let us focus on how it would be done. Look again at figure 6. On the right-hand side the plot for foreign cars is as we would wish for a back-to-back plot. What we need to change is the display on the left-hand side for domestic cars. Each line should be reversed and placed against the right-hand side. foreign is 0 for domestic cars (and 1 for foreign cars). Then the command

```
. by foreign mpg, sort: generate bbx = cond(foreign == 0, -_n, _n)
```

assigns negative x coordinates, -1 down, on the left-hand side and positive coordinates, 1 and up, on the right-hand side. (If you want a tutorial on cond(), see Kantor and Cox [2005].) We need to ensure that the largest value of the x coordinate appears in the left-hand panel and the smallest value in the right-hand panel. That way, the data points in the left-hand display nearly touch the right-hand edge of their display and vice versa. Some analysis shows that this goal is achieved by a translation (shunt, in plainer English) of the left-hand panel coordinates to the right:

```
. summarize bbx, meanonly
. replace bbx = bbx + max(-r(min), r(max)) + 1 if foreign == 0
```

Let us look at that again. By construction, the x coordinates are negative on the left-hand side and positive on the right-hand side. Thus r(min), which holds the minimum value of the x coordinates bbx after a summarize, is always negative. Its negation -r(min) is the length of the longest stem in the left-hand panel. r(max) is the length of the longest stem in the right-hand panel. Our translation is of whichever is larger, plus 1, to the right.

If you care about the detailed logic of that translation, focus on some examples. (Otherwise, skip to the next paragraph.) Suppose that the longest stems have length 4 on the left-hand side and length 3 on the right-hand side. These will have x coordinates $-4, -3, -2, -1$ and 1, 2, 3, respectively. For the displays to touch as desired, -4 must be translated to a number at least 1 and -1 must be translated to a number at least 3. The gentlest solution is the mapping from $-4, -3, -2, -1$ to 1, 2, 3, 4, namely, adding 5, the length of the longer stem plus 1. If the stems were reversed, namely, with x coordinates $-3, -2, -1$ and 1, 2, 3, 4, the same solution would apply. If we had two stems of equal length, say, $-3, -2, -1$ and 1, 2, 3, the same rule would apply: the longer stem (either one) has length 3, and adding 4 maps $-3, -2, -1$ to 1, 2, 3. We do not see any of these values, because the x-axis labels are all suppressed, but we do need to do this for the graph to appear as we would wish.

Once the translation is done, we have all we want, as shown in figure 8:

```
. scatter mpg bbx, by(foreign, compact) xla(none) xti("") ms(none) mla(mpg)
> mlabpos(0) yla(none)
```

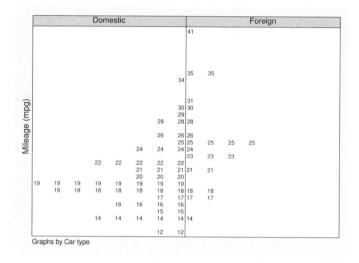

Figure 8: Back-to-back stem-and-leaf plots of miles per gallon for domestic and foreign cars

In typed back-to-back stem-and-leaf displays, stems are shown on a spine between the two parts. Having the stems as axis labels, whether shown or suppressed, is more straightforward when we do it graphically. A second set of labels could be shown on the right-hand axis if desired.

3.5   Other possibilities

By no means have we exhausted the possibilities for simple variations on the basic theme. Here are some more.

- *Reversing y-axis scale.* So far, as you will have noticed, we have followed the usual graph convention in which response magnitude, shown on the y axis, increases upward. The convention with text stem-and-leaf displays is the opposite. The graph option `ysc(reverse)` will produce that.

- *Different leaf choices.* Although we have considered in detail how to produce leaves from some or all of the digits of the variable whose distribution is being shown, doing so is not compulsory. Something different could be used for the leaf, including values coding for some categorical variable. The leaf is whatever is specified as marker label with the `mlabel()` option.

- *Swapping axes.* We can easily swap the y and x axes to produce a display in which magnitude is horizontal and leaves are stacked vertically.

- *Changing leaf size.* Being able to tweak the marker label size with the `mlabsize()` option is useful.

- *Aspect ratio.* We could also change the aspect ratio to use more or less horizontal space, often simply by tuning `xsize()`.

4   stemplot

One of my main aims in these columns is to emphasize how much you can do with a few basic commands to get where you want to be. Here being able to create or re-create basic graphs and to devise your own variations on them gives flexibility born from freedom. You do not depend on whether someone has written a program to do what you want to do or on whether you can find such a program if it exists. Nevertheless, producing even simple results can require mastery of fiddly little details: the manipulation needed to produce back-to-back displays is one example. Many users may not relish the minor acrobatics that may be needed.

Therefore, as an alternative, I publish a `stemplot` command with this column, encapsulating the main ideas so far, and then some more.

4.1   Syntax

`stemplot` *varname* [*if*] [*in*] [, `by`(*byvar* [, *byopts*]) `back` `digits`(*#*)
   `format`(*format*) `width`(*#*) `horizontal` *graph_options*]

`stemplot` produces stem-and-leaf plots and similar plots for numeric variable *varname* by using Stata's graphics. For stem-and-leaf plots as typed text displays, see [R] `stem` or `help stem`.

4.2   Options

`by`(*byvar* [, *byopts*]) specifies that displays are to be shown separately by groups defined by *byvar*. See [G-3] ***by_option*** or `help by_option`. The suboption `total` is legal but will not work as may be desired, because it superimposes, not juxtaposes, individual displays.

`back` specifies that a two-group display specified using `by()` be shown back to back.

`digits`(*#*) specifies the number of digits to be shown as leaves. The default is `digits(1)`. To be more precise, the rule for leaves is to show

   `substr(string(`*varname*`, "`*format*`"), -`*digits*`, `*digits*`)` or

string(*varname*, "*format*")

whenever that is empty. Here *format* is by default the display format of *varname*, as shown by describe and set by format. See also the format() option described below.

format(*format*) specifies a format other than the display format of *varname* to be shown in determining leaves. See also the digits() option described above. This option is seldom specified.

width(*#*) specifies the width of intervals defining distinct lines.

horizontal specifies a horizontal display in which the magnitude axis is horizontal and leaves are stacked vertically. back is not allowed with horizontal.

graph_options are any of the options documented in [G-2] **graph twoway scatter**. For example, mlabel() overrides the program's choice of leaf, and mlabsize() controls the size of leaves.

4.3   Comments

How does the scope of stemplot differ from that of combinations of individual commands, as discussed previously? Here we note the most important features.

1. stemplot preserves sort order, as is standard programming practice for any program not intended to change that order.

2. There are more checks on user input.

3. stemplot allows specifying if and in.

4. Various options allow tuning of the display.

4.4   A last look at those cars

Suppose that we wanted to show some information about the make of cars. The full name within the variable make is too long to fit comfortably on a stem-and-leaf display, but we can compromise on the first word:

```
. generate M = word(make, 1)
```

Some experimentation shows that we should sort within stems, use a different choice of mlabpos(), and stretch the *x*-axis scale so that the rightmost marker label fits:

```
. sort mpg M
. stemplot mpg, mla(M) mlabpos(3) xsc(r(. 10))
```

To save space, I do not show the result here, but you can experiment for yourself.

5   Choral finale

For a final example, we leave the world of automobiles and turn to choral music. Chambers, Cleveland, Kleiner, and Tukey (1983, 350) listed the heights of singers in the New York Choral Society in 1979. Cleveland (1993) plotted the dataset in various ways; it is downloadable from http://www.stat.purdue.edu/~wsc/visualizing.tables.txt.

Here we focus first on the broad subdivision between soprano, alto, tenor, and bass parts, ignoring the distinction between first (higher pitch) and second (lower pitch) groups for each part.

A general relationship between singer part and height is no surprise. What a stem-and-leaf plot can show is the fine structure of the distributions, based on the singers' own statements. Here the choice of units is important, because of their psychological overtones. Although the data are reported in inches, people in countries like the United States commonly think of their heights in feet and inches, not just in inches. An adult height of 6 feet (72 inches) is moderately tall by most standards (although clearly not, for example, that of basketball players). Similarly, an adult height of 5 feet (60 inches) is rather short by most standards, or petite if you prefer. Fully metric readers might note that 6 feet is about 1.83 meters and 5 feet about 1.52 meters.

Ordinary experience shows that although many adults are happy with their heights, a few would prefer to be taller and a few, shorter. Also, some people may be self-conscious about being distinct from whatever they regard as a norm. Let us bear all this in mind while examining these data.

Using `stemplot` is convenient here to produce figure 9.

```
. stemplot height, by(spart, row(1) compact noiyitic
> note(units are inches, place(e))) yla(60 "5´" 66 "5´ 6´´" 72 "6´")
> yaxis(1 2) yla(60(6)72, axis(2) ang(h))
```

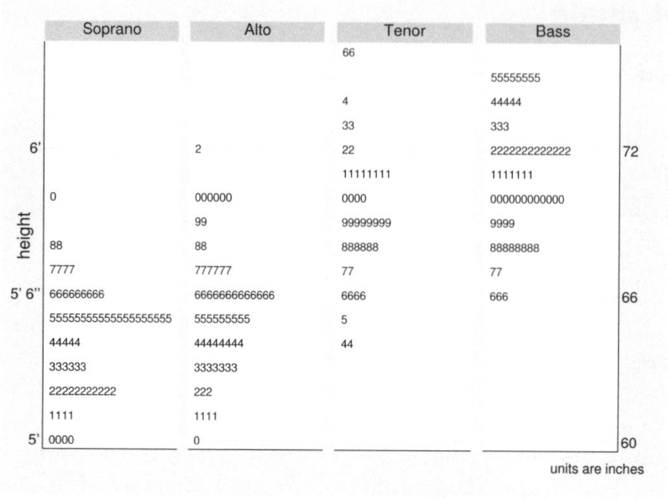

Figure 9: New York Choral Society singers' heights, 1979. Note not only the general relationship between singer part and height but also intriguing hints of fine structure.

spart is a numeric variable with value labels, which we need here to show soprano, alto, tenor, and bass in their natural order. A string variable would sort alphabetically from alto to tenor, which would not be a good choice. Two y axes permit two sets of labels, one in feet and inches and one in inches, making the comparison easier, even though all readers can presumably divide small numbers by 12.

The plot does indeed show intriguing, although contradictory, fine structure. No singers admit to being less than 60 inches (5 feet) tall. The modes for sopranos at 55 inches and for altos at 56 inches may be genuinely different, or they may reflect some personal preferences. Why do no altos report themselves as 71 inches? Is there some wishful thinking among some of the basses who claim 72 inches (6 feet)? There are hints of other features that may owe more to psychology than to anatomy.

We can also bring the split into higher and lower groups for each part (first and second sopranos, and so forth). One way to explore that is to use that variable as a new leaf (figure 10).

```
. stemplot height, by(spart, row(1) noiytic
> note(1 = high; 2 = low, place(e))) mla(group)
> yla(60 "5´" 66 "5´ 6´´" 72 "6´") yaxis(1 2) yla(60(6)72, axis(2) ang(h))
```

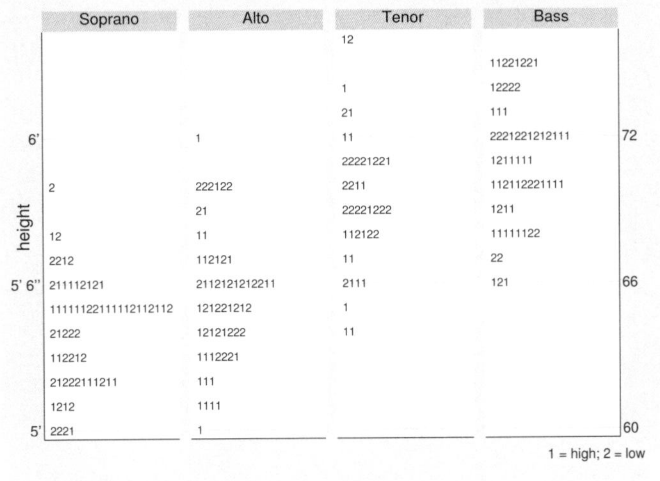

Figure 10: New York Choral Society singers' heights, 1979, by singer part and groups with relatively high or low pitch

Naturally, any fine structure identified in this way should be tested by independent evidence. We need to be clear that we are not seizing on quirks produced by sampling variation. As Harrell (2001, ix) aptly remarks, "Using the data to guide the data analysis is almost as dangerous as not doing so." The opportunity for measuring these singers again is long past, but there may be similar (preferably larger) datasets that could provide some checks. Statistical and scientific caution aside, the stem-and-leaf plot is one of the graphical tools to try when fine structure within distributions is of interest.

6   Conclusion

Stem-and-leaf displays are one of the staples of exploratory statistical graphics. However, producing them as text displays, as with Stata's `stem` command, has inevitable limitations, especially for comparison of distributions across groups. Revisiting stem-and-leaf as full-blown graphs requires no more than the idea of scatterplots with marker labels. A few steps lead not only to acceptable stem-and-leaf displays but also to variants on them. Comparison of groups is then easy by using the standard `by()` option of `twoway scatter`.

This column also illustrates a useful Stata strategy of using basic commands and toy examples and moving step by step toward what you want. Technique is thus strengthened. Serendipitous discoveries of unforeseen possibilities are likely. Although the project followed through to produce a new `stemplot` command, the greater lesson is the flexibility and power of the language available to you.

7 References

Chambers, J. M., W. S. Cleveland, B. Kleiner, and P. A. Tukey. 1983. *Graphical Methods for Data Analysis*. Belmont, CA: Wadsworth.

Cleveland, W. S. 1993. *Visualizing Data*. Summit, NJ: Hobart.

Cox, N. J. 2002. Speaking Stata: How to move step by: step. *Stata Journal* 2: 86–102.

Darton, M., and J. Clark. 1994. *The Dent Dictionary of Measurement*. London: Dent.

Dudley, J. W. 1946. *Examination of Industrial Measurements*. New York: McGraw–Hill.

Emerson, J. D., and D. C. Hoaglin. 1983. Stem-and-leaf displays. In *Understanding Robust and Exploratory Data Analysis*, ed. D. C. Hoaglin, F. Mosteller, and J. W. Tukey, 7–32. New York: Wiley.

Griffiths, D., W. D. Stirling, and K. L. Weldon. 1998. *Understanding Data: Principles and Practice of Statistics*. Brisbane: Wiley.

Harrell, F. E., Jr. 2001. *Regression Modeling Strategies: With Applications to Linear Models, Logistic Regression, and Survival Analysis*. New York: Springer.

Hoaglin, D. C. 1983. Letter values: A set of selected order statistics. In *Understanding Robust and Exploratory Data Analysis*, ed. D. C. Hoaglin, F. Mosteller, and J. W. Tukey, 33–57. New York: Wiley.

Kantor, D., and N. J. Cox. 2005. Depending on conditions: A tutorial on the cond() function. *Stata Journal* 5: 413–420.

Moore, D. S., and G. P. McCabe. 2006. *Introduction to the Practice of Statistics*. New York: Freeman.

Pennycuick, C. J. 1988. *Conversion Factors: SI Units and Many Others*. Chicago: University of Chicago Press.

Preece, D. A. 1981. Distribution of final digits in data. *Statistician* 30: 31–60.

Tufte, E. R. 1990. *Envisioning Information*. Cheshire, CT: Graphics Press.

Tukey, J. W. 1972. Some graphic and semigraphic displays. In *Statistical Papers in Honor of George W. Snedecor*, ed. T. A. Bancroft and S. A. Brown, 293–316. Ames, IA: Iowa State University Press.

———. 1977. *Exploratory Data Analysis*. Reading, MA: Addison–Wesley.

Wild, C. J., and G. A. F. Seber. 2000. *Chance Encounters: A First Course in Data Analysis and Inference*. New York: Wiley.

About the author

Nicholas Cox is a statistically minded geographer at Durham University. He contributes talks, postings, FAQs, and programs to the Stata user community. He has also coauthored 15 commands in official Stata. He wrote several inserts in the *Stata Technical Bulletin* and is an editor of the *Stata Journal*.

The Stata Journal (2008)
8, Number 1, pp. 105–121

Speaking Stata: Spineplots and their kin

Nicholas J. Cox
Department of Geography
Durham University
Durham City, UK
n.j.cox@durham.ac.uk

Abstract. The term *spineplot* has been applied over the last decade or so to a type of bar chart used particularly for showing frequencies, proportions, or percentages of two cross-classified categorical variables. The principle is that the areas of rectangular tiles are proportional to the frequencies in the cells of a contingency table. Often both coarse and fine structure are easy to see, including departures from independence. The main idea has, in fact, been rediscovered repeatedly over at least the last 130 years. In its most general form, it has been widely publicized under the name *mosaic plots*. This column introduces, discusses, and exemplifies a Stata implementation of spineplots. It is noted that a restriction to two variables is more apparent than real, as either axis of a spineplot can show a composite variable defined by cross combinations of two or more variables.

Keywords: gr0031, spineplots, mosaic plots, bar charts, graphics, categorical data

1 Introduction

The recent history of categorical data analysis within statistical science has been marked by increasing convergence with what might reasonably be dubbed continuous data analysis. Even a generation ago, categorical data analysis was little more to practitioners in many quantitative fields than a ragbag of chi-squared tests and measures of bivariate association. (Indeed, even now many introductory texts appear to offer little more.) In stark contrast, those looking at continuous response variables could exploit a steadily more coherent and powerful toolbox based on regression and ANOVA, seen as members of a family of linear models. However, much greater focus in categorical data analysis over the last few decades on models of various kinds, including log linear and logit models and their several relatives, has greatly lessened the contrasts between the two major parts of statistical practice (see, for example, Agresti [2002]).

One facet of categorical data analysis which continues to receive uneven attention is the use of graphical methods. It is often argued (for example, by Tufte [2001]) that tabular displays, whether of data or summaries or model results, may be more effective or informative than graphs for many categorical problems. Nevertheless, various plotting methods have been suggested for such problems. Friendly (2000) surveys many recent innovations, but none yet appear to challenge bar charts as the most popular graphical method for categorical data.

Bar charts provoke a range of reactions from statistically minded people. Some charts showing only a few frequencies may strike readers as a waste of space in any outlet supposedly aimed at intelligent adults or as too elementary or trivial to deserve much coverage in professional literature. Yet there are many reasons for thinking that bar charts may complement tables helpfully, particularly when the bar charts are well designed and well chosen.

In a previous column, I reviewed some ways of producing such charts in Stata for categorical data (Cox 2004). In this column, I focus on what are now widely known as *spineplots*, discussing the main ideas of spineplots and showing a Stata implementation. The term may be new to you, but the idea may yet be familiar; in any case, it will not appear strange. Spineplots grow out of the basic graphical notion that area may usefully encode frequency, which underlies several other standard forms, including histograms.

2   Spineplots

Names should not matter, but they do. Labels should matter much less than the underlying ideas. A wind rose or a stem-and-leaf plot by any other name is just as sweet, or as prickly, an idea. Yet across times and places and disciplines, all sorts of minor and major confusions can arise when the same name is used for different things, different names are used for the same thing, or authors unthinkingly assume that readers have had the same education and experience and possess the same terminology. Explaining what is, and what is not, a spineplot—or more precisely what is and is not done by the Stata program `spineplot`—thus requires attention to usages in the literature.

The name *spineplot* is credited to Hummel (1996). The term is gaining in popularity but appears already to be differently understood. In the strictest definition, spineplots are one-dimensional, horizontal stacked bar charts, but many discussions and implementations allow vertical subdivision (e.g., by highlighting) into two or possibly more categories. Some literature treats spineplots, as understood here, under the heading of *mosaic plots* (or *mosaicplots*), variously with and without also using the term *spineplot*.

The Stata implementation `spineplot` discussed here adopts a broad interpretation of the term. It works on two categorical variables—not one—and conveys the frequencies shown in a two-way contingency table. (One-dimensional, horizontal stacked bar charts have long been possible in Stata; in Stata 8 the official command `graph hbar` became available.) Conversely, the implementation here does not purport to be a general mosaic plot program capable of producing mosaic plots given three or more categorical variables.

Textbooks and monographs with examples of spineplots and related plots include Friendly (2000); Venables and Ripley (2002); Robbins (2005); Unwin, Theus, and Hofmann (2006); Young, Valero-Mora, and Friendly (2006); and Cook and Swayne (2007). Among several papers, Hofmann's (2000) discussion is clear, concise, and well illustrated.

Mosaic plots, including spineplots as a special case, have been reinvented several times under different names. Hartigan and Kleiner (1981, 1984) introduced them, or reintroduced them, into mainstream statistics. Friendly (2002) cites earlier examples, including the work of Georg von Mayr (1877), Karl G. Karsten (1923), and Erwin J.

Raisz (1934). Hofmann (2007) discusses a mosaic by Francis A. Walker (1874). Other early examples are those of Willard C. Brinton (1914, quoting earlier work), Berend G. Escher (1924), and Hans Zeisel (1947, 1985).[1] Further, independent reinventions of the idea continue to appear (e.g., Bertin [1983]; Feinstein and Kwoh [1988]; and Feinstein [2002]).

3 First examples

Examples will convey the essence far better than a word description. With a nod to Stata tradition, fire up Stata with the `auto` data, and look at the cross classification of two categorical variables: whether cars are foreign (from outside the United States) `foreign` and their 1978 repair record `rep78`. Repair record may be considered to be a response variable; hence, as with scatter plots and the Stata command `scatter`, it is named first to `spineplot` as the variable to be shown on the *y* axis. `spineplot` does not try to be smart about colors, nor does it know whether a categorical variable is ordered (ordinal) or not (nominal). Thus we here skip the default and move directly to specifying an ordered series of gray scales for bar colors (figure 1):

```
. sysuse auto
(1978 Automobile Data)

. spineplot rep78 foreign, bar1(bcolor(gs14)) bar2(bcolor(gs11))
> bar3(bcolor(gs8)) bar4(bcolor(gs5)) bar5(bcolor(gs2))
```

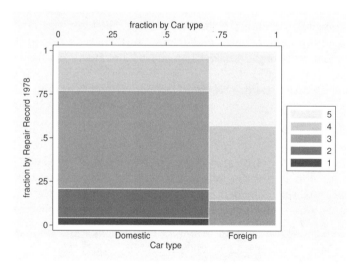

Figure 1: Spineplot of repair record and whether foreign for 74 cars, as produced by `spineplot`

---

1. See Anonymous (1967), Robinson (1970), Sills (1992), Anderson (2001), and Hertz (2001) for biographical pieces on several of these pioneers. Karl Karsten has been credited with the idea of hedge funds. Berend Escher is now better known as a brother of Maurits C. Escher, whose own mosaics are immensely more intricate and intriguing than any to be discussed here.

As you might guess, options like `bar1()` and `bar2()` override defaults for the first, second, and subsequent bars. Counting is from the top downward. Here the darkest gray scales show poor repair records. Adopting the reverse choice, or indeed any other choice of colors, is naturally at your discretion. Whatever the choice, the spineplot makes clear that foreign and domestic cars had very different distributions of repair record in 1978.

The graph structure is similar to the structure of a standard two-way contingency table, such as the one `tabulate rep78 foreign` would produce. One detailed difference is that high response values are in the last rows but toward the top of the y axis, reflecting table and graph conventions, respectively. Another detailed difference is that cells with zero frequency are represented in the spineplot by tiles of zero area, that is, not at all.

For interpretation of spineplots, note that cross classification of independent variables would yield tiles that align consistently, as the resulting conditional distributions would be identical. Conversely, departures from independence, or relationships between variables, are shown by failure of alignment. The fine structure of such departures is open to inspection, although limits are imposed by the low visibility of cells with low frequencies and thus low tile areas. Spineplots are especially useful when considering a null hypothesis of independence.

However, in some cases where independence is highly implausible, spineplots may not be particularly effective. A common example is assessing categorical agreement of observers or methods, the problem which to many users is that addressed by the `kappa` command ([R] **kappa**). Here the usual expectation is that the diagonal or near-diagonal cells of the contingency table would show much higher frequencies than those near the opposite corners. Such a pattern would indeed be obvious on a spineplot, but the coloring used in `spineplot` does not make further scrutiny especially helpful.

Be that as it may, let us consider how this spineplot differs from more conventional bar charts. Surprising although it may seem, official Stata offers no direct and obvious command for bar charts of categorical data. Two user-written commands, `catplot` and `tabplot`, are among the alternatives (Cox 2004). Both may be downloaded from the Statistical Software Components archive by using the `ssc` command (see [R] **ssc** for further information).

With `catplot`, there is considerable choice of format. Two close relatives of the spineplot are particularly pertinent. The first shows frequencies (figure 2):

```
. tabulate rep78 foreign
  (output omitted)
. catplot bar rep78 foreign, asyvars stack bar(1, bcolor(gs2))
> bar(2, bcolor(gs5)) bar(3, bcolor(gs8)) bar(4, bcolor(gs11))
> bar(5, bcolor(gs14)) legend(pos(3) col(1))
```

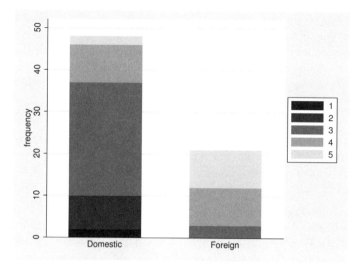

Figure 2: Bar chart of repair record and whether foreign for 74 cars, as produced by `catplot`

The second shows stacked percentages (figure 3):

```
. catplot bar rep78 foreign, asyvars stack percent(foreign) bar(1, bcolor(gs2))
> bar(2, bcolor(gs5)) bar(3, bcolor(gs8)) bar(4, bcolor(gs11))
> bar(5, bcolor(gs14)) legend(pos(3) col(1))
```

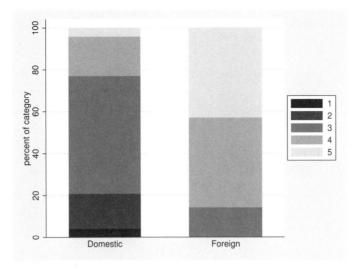

Figure 3: Bar chart of repair record and whether foreign for 74 cars, as produced by `catplot`, showing column percentages

With `catplot`, therefore, as with most bar chart software, it is easy to get a display of stacked frequencies. In that display, proportions or percentages are tacit and so often difficult to read off precisely. It is also easy to get a display of stacked percentages. In that display, the underlying frequencies are not in view. (In this case, `catplot` is a wrapper for `graph bar`, which might suggest the use of the `blabel()` option. But `blabel()` shows numerically what is being shown graphically, and we would want to show something else, so `blabel()` would not help.)

`tabplot` is another possibility. Here the percentage breakdown is shown in figure 4. Omitting the `percent()` option would yield a display of frequencies instead.

```
. tabplot rep78 foreign, percent(foreign) showval(format(%2.1f))
```

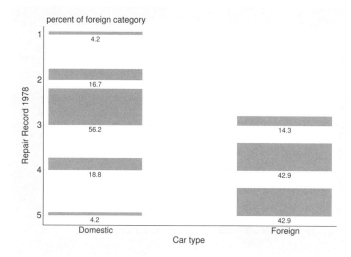

Figure 4: Tabular bar chart of repair record and whether foreign for 74 cars, as produced by `tabplot`, showing column percentages

This plot echoes the structure of a two-way contingency table even more clearly than does a spineplot. A glance at the code shows that much of the work within `tabplot` is done by a call to `twoway rbar`. But again there is a choice between showing frequencies and showing percentages. There is no scope for showing both simultaneously.

In sum: Spineplots show conditional distributions on both axes simultaneously. We can easily add information on absolute frequencies using the `text()` option (figure 5):

```
. by foreign rep78, sort: generate N = _N
. spineplot rep78 foreign, bar1(bcolor(gs14)) bar2(bcolor(gs11))
> bar3(bcolor(gs8)) bar4(bcolor(gs5)) bar5(bcolor(gs2)) text(N)
```

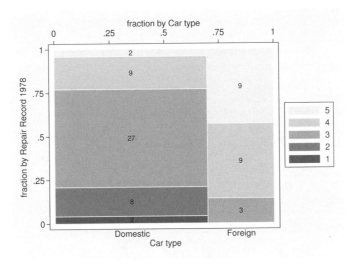

Figure 5: Spineplot of repair record and whether foreign for 74 cars, as produced by `spineplot`, with cell frequencies shown

Missing values in either of the two variables do not perturb the frequencies produced by the `generate` command above. The resulting frequencies are assigned but then ignored by `spineplot`. Conversely, empty cells of the contingency table do not, by definition, correspond to any observations, so counts of zero will not be shown. Combining the count with an `if` or `in` condition would require more care, but the details need not detain us now. Plotting something else, such as standardized residuals given some model, is another possibility. It would often be a good idea to impose a particular numeric format before display, say, by `string(`*residual*`, "%4.3f")`.

Most implementations of spineplots (and, more generally, mosaic plots) in other software omit axes and numerical scales and convey a recursive subdivision according to what may be several categorical variables by a hierarchy of gaps of various sizes. As the graphs produced by `spineplot` are restricted to two variables, axes and numerical scales are kept as defaults. The distinction between categories is conveyed by bar boundaries rather than explicit gaps. Naturally, there is scope for omitting graph elements not desired using standard `graph` options, or, in Stata 10 upward, the Graph Editor. Similarly, users may vary the thickness of bar boundaries, although thick boundaries would distort the relative sizes of what are perceived as bar areas.

The examples already seen raise other small matters of presentation.

First, note the possibility of using `plotregion(margin(zero))` to place axes alongside the plot region. Having a margin is often useful for scatterplots and their kin but is perhaps distracting for spineplots.

As with scatterplots, response variables are usually better shown on the y axis of spineplots. But as with scatter plots, there can be reasons for overriding that convention. (In the Earth or environmental sciences, plotting height above or depth below the land surface on the vertical axis is common and indeed often expected.) If one variable is binary, it is often better to plot that one on the y axis. The `foreign` variable is a case in point. Even though `foreign` is arguably a predictor of `rep78` rather than vice versa, I suggest that the spineplot with `foreign` on the y axis is more congenial. See figure 6 and judge for yourself. Notice that ordering of colors is now less of an issue, as any two distinct colors are ordered one way or the other.

```
. spineplot foreign rep78
```

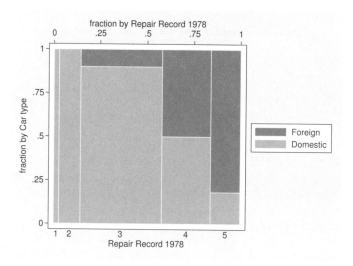

Figure 6: Spineplot of whether foreign and repair record for 74 cars, as produced by `spineplot`, with cell frequencies shown

Even more mundane, but very possibly troublesome in practice, is that if one or more cells have very small frequencies, then a squeeze of some sort is inevitable with `spineplot`. There is no way to show the corresponding tiles, or descriptive labels, or added text, without some difficulty. There are no easy solutions to this problem. You may decide to amalgamate cells; or to use the Graph Editor to ease crowding by moving text, adding arrows, and so forth; or just to use some other kind of graph. Manifestly, all kinds of graphs have some limitations on what they can show easily and effectively, and spineplots are no exception.

4   Discrimination at Berkeley?

A now classic problem among categorical analysts concerns the success or failure of applications for admittance as graduate students at the University of California, Berkeley. The problem was first discussed by Bickel, Hammel, and O'Connell (1975) and since then worked over in various ways in many articles and texts (e.g., Freedman, Pisani, and Purves [1978; 2007]; Friendly [2000]; and Agresti [2002]). Here we use a subset of the data presented by Friendly (2000) and Agresti (2002). The response is decision—admitted or rejected—and the covariates are intended major (masked by identifiers A, B, C, D, E, F) and sex of applicants. The data are available with the files for this column as `berkeley.dta`. They come as frequencies of the various cross combinations, so we must specify weights when we call up `spineplot`. (Alternatively, `expand`ing the dataset on the frequencies so that every individual application became an observation would make that unnecessary; see [D] **expand** for more.)

```
. use berkeley, clear
. spineplot decision sex [fw=frequency]
```

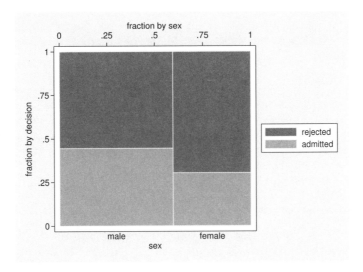

Figure 7: Spineplot of decision versus sex for admissions to various Berkeley graduate majors. At first sight, substantial discrimination against females is evident.

A spineplot of decision versus gender shows apparent discrimination against females (figure 7). However, majors are by no means equally easy to get into (figure 8). A corresponding `tabulate` shows that admission rates vary from 64% for A to 6.4% for F.

```
. spineplot decision major [fw=frequency]
```

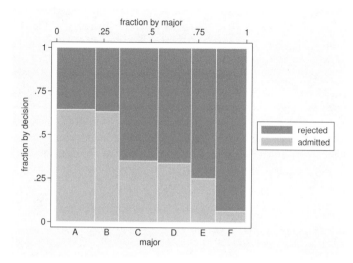

Figure 8: Spineplot of decision versus major for admissions to various Berkeley graduate majors. Acceptance rates vary over a tenfold range.

These are just two-dimensional representations of three-dimensional data. We need to see what structure may exist in three dimensions, including whether there are interactions between the covariates. How can we do that with a two-dimensional display? The answer lies in a composite categorical variable, defined by the cross combinations of two or more categorical variables (Cox 2007). Although not the only method, `egen`'s `group()` function is fine for this purpose:

```
. egen group = group(major sex), label
```

The `label` option is essential for graphs and tables to make sense. Without it, the resulting groups would just show as groups 1 to 12. Further, the order of variables fed to the function is crucial. `group(major sex)` aligns male and female for each major. `group(sex major)` would align majors for each sex. The first is what we need here. In other problems, experimentation with group order may be needed to see what works best.

```
. spineplot decision group [fw=frequency], xlabel(, angle(v) axis(2))
> xtitle("", axis(2)) xtitle(fraction by major and sex, axis(1))
```

Figure 9 shows the result. Vertical axis labels are the lesser of two evils, as there is far too little room for horizontal labels to be legible. Some readers may prefer to try a compromise, say, an angle of $45°$. The default title for the bottom x axis would be the variable label for `group`, `group(major sex)`, which we prefer to blank out. Similarly, the title for the top x axis improves on the default.

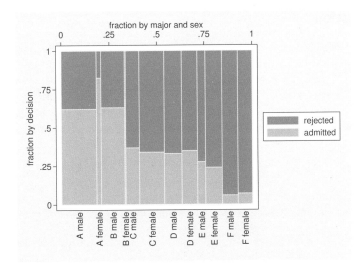

Figure 9: Spineplot of decision versus major and sex for admissions to various Berkeley graduate majors. Females are admitted proportionally more than males to four majors and proportionally less to the remaining two.

The fine structure of the display allows focus on the key question. Major by major, a higher proportion of females than males is admitted to A, B, D, and F, and a lower proportion to C and E. (Admittedly, the comparison for B is not clear on the graph given the small frequencies concerned; for that result, a peek at a table is needed.) Hence, the appearance of discrimination against females appears very much an artifact of the sex and major composition of the applicants or, in other terminology, an example of the amalgamation paradox often named for E. H. Simpson, despite its earlier elucidation by G. U. Yule and several others (Agresti 2002).

A lesson for other examples is that the restriction of spineplots to two variables is more apparent than real given the scope for creating composite variables. Compare what Hofmann (2001) calls "double-decker plots" (for binary responses) and what Wilkinson (2005) calls "region trees".

5   Spineplot details

5.1   Syntax

spineplot *yvar xvar* [*if*] [*in*] [*weight*] [,

    bar1(*twoway_bar_options*) ... bar20(*twoway_bar_options*)

    barall(*twoway_bar_options*) <u>miss</u>ing <u>percent</u>

    text(*textvar* [, *marker_label_options*]) *twoway_options*]

<u>fweight</u>s and <u>aweight</u>s may be specified; see [U] **11.1.6 weight**.

5.2   Description

`spineplot` produces a spineplot for two-way categorical data. The fractional breakdown of the categories of the first-named variable *yvar* is shown for each category of
the second-named variable *xvar*. Stacked bars are drawn with vertical extent showing
fraction in each *yvar* category given each *xvar* category and horizontal extent showing
fraction in each *xvar* category. Thus the areas of tiles formed represent the frequencies,
or more generally totals, for each cross combination of *yvar* and *xvar*.

5.3   Options

`bar1`(*twoway_bar_options*) ... `bar20`(*twoway_bar_options*) allow specification of the
appearance of the bars for each category of *yvar* using options of `twoway bar`.

`barall`(*twoway_bar_options*) allows specification of the appearance of the bars for all
categories of *yvar* using options of `twoway bar`.

`missing` specifies that any missing values of either of the variables specified should also
be included within their own categories. The default is to omit them.

`percent` specifies labeling as percentages. The default is labeling as fractions.

`text`(*textvar* [, *marker_label_options*]) specifies a variable to be shown as text at the
center of each tile. *textvar* may be a numeric or string variable. It should contain
identical values for all observations in each cross combination of *yvar* and *xvar*. A
simple example is the frequency of each cross combination. To show nothing in
particular tiles, use a variable with missing values (either numeric missing or empty
strings) for those tiles. A numeric variable with fractional part will typically look
best converted to string as, for example, `string`(*residual*,`"%4.3f"`). The user is
responsible for choice of tile colors so that text is readable. `text()` may also include
marker_label_options for tuning the display.

twoway_options refers to options of `twoway`; see [G-3] ***twoway_options***. By default
there are two *x* axes, `axis(1)` on top and `axis(2)` on bottom, and two *y* axes,
`axis(1)` on right and `axis(2)` on left.

5.4   Inside the program

You may wish to know more about how the program works. The code, naturally, is
open for inspection in your favorite text editor.

The program works by calculating cumulative frequencies. The plot is then produced
by overlaying distinct graphs, each being a call to `twoway bar, bartype(spanning)` for
one category of *yvar*. By default, each bar is shown with `blcolor(bg) blw(medium)`,
which should be sufficient to outline each bar distinctly but delicately. By default also,
the categories of *yvar* will be distinguished according to the graph scheme you are using.
With the default `s2color` scheme, the effect is reminiscent of canned fruit salad (which

may be fine for exploratory work). For a publishable graph, you might want to use something more subdued, such as various gray scales or different intensities, as in this column.

Options `bar1()` to `bar20()` are provided to allow overriding the defaults on up to 20 categories, the first, second, etc., shown. The limit of 20 is plucked out of the air as more than any user should really want. The option `barall()` is available to override the defaults for all bars. Any `bar#()` option always overrides `barall()`. Thus if you wanted thicker `blwidth()` on all bars, you could specify `barall(blwidth(thick))`. If you wanted to highlight the first category only, you could specify `bar1(blwidth(thick))` or a particular color.

Other defaults include `legend(col(1) pos(3))`. At least with `s2color`, a legend on the right implies an approximately square plot region, which can look quite good. A legend is supplied partly because there is no guarantee that all *yvar* categories will be represented for extreme categories of *xvar*. However, it will often be possible and tasteful to omit the legend and show categories as axis label text.

6   Conclusion

Spineplots offer an alternative to more conventional bar charts for showing the data in a two-way contingency table. Their particular merit arises from the fact that frequencies are encoded by tile areas so that, in principle, spineplots convey the information in both marginal and conditional distributions. Departure from independence is shown by failure of tiles to align, which is easily seen. Spineplots can also be extended to higher-order contingency tables, in so far as two or more categorical variables may be combined to form a single composite variable to be shown on either axis.

However, what is a key feature of spineplots can also be a limitation. Cells with small frequencies will be represented by small tiles, and cells with zero frequencies will not be represented at all, so the fine structure associated with such cells may be difficult to discern. Hence, other kinds of bar charts remain complementary for showing the structure of contingency tables.

7   Acknowledgments

Matthias Schonlau, Scott Merryman, and Maarten Buis provoked the writing of the `spineplot` command through challenging Statalist postings. A suggestion from Peter Jepsen led to the `text()` option. Private emails from Matthias Schonlau and Antony Unwin highlighted different senses of spineplots and the importance of sort order. Antony suggested standardizing on "spineplot" rather than "spine plot". Maarten verified for me that the spineplot in my copy of Escher (1934) also appears in Escher (1924). Vince Wiggins originally told me about the undocumented `bartype(spanning)` option.

8   References

Agresti, A. 2002. *Categorical Data Analysis*. 2nd ed. Hoboken, NJ: Wiley.

Anderson, M. J. 2001. Francis Amasa Walker. In *Statisticians of the Centuries*, ed. C. C. Heyde and E. Seneta, 216–218. New York: Springer.

Anonymous. 1967. In memoriam Prof. Dr. B. G. Escher. *Geologie en Mijnbouw* 46: 417–422.

Bertin, J. 1983. *Semiology of Graphics: Diagrams, Networks, Maps*. Madison: University of Wisconsin Press.

Bickel, P. J., E. A. Hammel, and J. W. O'Connell. 1975. Sex bias in graduate admissions: Data from Berkeley. *Science* 187: 398–404.

Brinton, W. C. 1914. *Graphic Methods for Presenting Facts*. New York: Engineering Magazine Company.

Cook, D., and D. F. Swayne. 2007. *Interactive and Dynamic Graphics for Data Analysis: With R and GGobi*. New York: Springer.

Cox, N. J. 2004. Speaking Stata: Graphing categorical and compositional data. *Stata Journal* 4: 190–215.

———. 2007. Stata tip 52: Generating composite categorical variables. *Stata Journal* 7: 582–583.

Escher, B. G. 1924. *De Methodes der Grafische Voorstelling*. Amsterdam: Wereldbibliotheek.

———. 1934. *De Methodes der Grafische Voorstelling*. 2nd ed. Amsterdam: Wereldbibliotheek.

Feinstein, A. R. 2002. *Principles of Medical Statistics*. Boca Raton, FL: Chapman & Hall/CRC.

Feinstein, A. R., and C. K. Kwoh. 1988. A box-graph method for illustrating relative size relationships in a 2×2 table. *International Journal of Epidemiology* 17: 222–224.

Freedman, D., R. Pisani, and R. Purves. 1978. *Statistics*. New York: W. W. Norton.

———. 2007. *Statistics*. 4th ed. New York: W. W. Norton.

Friendly, M. 2000. *Visualizing Categorical Data*. Cary, NC: SAS Institute.

———. 2002. A brief history of the mosaic display. *Journal of Computational and Graphical Statistics* 11: 89–107.

Hartigan, J. A., and B. Kleiner. 1981. Mosaics for contingency tables. In *Computer Science and Statistics: Proceedings of the 13th Symposium on the Interface*, ed. W. F. Eddy, 268–273. New York: Springer.

————. 1984. A mosaic of television ratings. *American Statistician* 38: 32–35.

Hertz, S. 2001. Georg von Mayr. In *Statisticians of the Centuries*, ed. C. C. Heyde and E. Seneta, 219–222. New York: Springer.

Hofmann, H. 2000. Exploring categorical data: Interactive mosaic plots. *Metrika* 51: 11–26.

————. 2001. Generalized odds ratios for visual modeling. *Journal of Computational and Graphical Statistics* 10: 628–640.

————. 2007. Interview with a centennial chart. *Chance* 20(2): 26–35.

Hummel, J. 1996. Linked bar charts: Analysing categorical data graphically. *Computational Statistics* 11: 23–33.

Karsten, K. G. 1923. *Charts and Graphs: An Introduction to Graphic Methods in the Control and Analysis of Statistics*. New York: Prentice Hall.

Raisz, E. J. 1934. The rectangular statistical cartogram. *Geographical Review* 24: 292–296.

Robbins, N. M. 2005. *Creating More Effective Graphs*. Hoboken, NJ: Wiley.

Robinson, A. H. 1970. Erwin Josephus Raisz, 1893–1968. *Annals of the Association of American Geographers* 60: 189–193.

Sills, D. L. 1992. In memoriam: Hans Zeisel, 1905–1992. *Public Opinion Quarterly* 56: 536–537.

Tufte, E. R. 2001. *The Visual Display of Quantitative Information*. 2nd ed. Cheshire, CT: Graphics Press.

Unwin, A., M. Theus, and H. Hofmann. 2006. *Graphics of Large Datasets: Visualizing a Million*. New York: Springer.

Venables, W. N., and B. D. Ripley. 2002. *Modern Applied Statistics with S*. 4th ed. New York: Springer.

von Mayr, G. 1877. *Die Gesetzmässigkeit im Gesellschaftsleben*. München: Oldenbourg.

Walker, F. A. 1874. *Statistical Atlas of the United States Based on the Results of the Ninth Census 1870*. New York: Census Office.

Wilkinson, L. 2005. *The Grammar of Graphics*. 2nd ed. New York: Springer.

Young, F. W., P. M. Valero-Mora, and M. Friendly. 2006. *Visual Statistics: Seeing Data with Dynamic Interactive Graphics*. Hoboken, NJ: Wiley.

Zeisel, H. 1947. *Say It with Figures*. New York: Harper.

————. 1985. *Say It with Figures*. 6th ed. New York: Harper & Row.

About the author

Nicholas Cox is a statistically minded geographer at Durham University. He contributes talks, postings, FAQs, and programs to the Stata user community. He has also coauthored 15 commands in official Stata. He wrote several inserts in the *Stata Technical Bulletin* and is an editor of the *Stata Journal*.

The Stata Journal (2008)
8, Number 2, pp. 269–289

Speaking Stata: Between tables and graphs

Nicholas J. Cox
Department of Geography
Durham University
Durham City, UK
n.j.cox@durham.ac.uk

Abstract. Table-like graphs can be interesting, useful, and even mildly innovative. This column outlines some Stata techniques for producing such graphs. `graph dot` is likely to be the most under-appreciated command among all existing commands. Using `by()` with various choices is a good way to mimic a categorical axis in many `graph` commands. When `graph bar` or `graph dot` is not flexible enough to do what you want, moving to the more flexible `twoway` is usually advisable. `labmask` and `seqvar` are introduced as new commands useful for preparing axis labels and axis positions for categorical variables. Applications of these ideas to, e.g., confidence interval plots lies ahead.

Keywords: gr0034, labmask, seqvar, tables, graphs, dot charts

1 Introduction

When reporting on data analyses, researchers may need to choose between tables and graphs. The choice may be severely constrained. An advisor, a supervisor, a reviewer, or an editor may lay down instructions about acceptable forms of reporting. Many disciplines have firm conventions on the format of tables of parameter estimates, t statistics, p-values, and the like, or on ways of graphing results. Yet often there is much room for discussion on whether tables or graphs may better serve a particular aim and a particular audience. Tufte (2001, 56) suggested that "Tables usually outperform graphics in reporting on small data sets of 20 numbers or less". Gelman, Pasarica, and Dodhia (2002) in contrast emphasized the scope for turning tables into graphs. This column focuses on the transition zone between tables and graphs by discussing some Stata techniques for graphs with table-like structure or content.

Graphs may resemble tables if axes show sets of categories rather than numeric scales. Stata's `graph` command has a formal idea of a categorical axis, which is embedded in the commands `graph dot`, `graph bar`, and `graph box`. These commands show values for a numeric variable according to the groups of one or more categorical variables. Their common structure is that a response variable is considered to be plotted on the y axis—even if that is horizontal, as with `graph hbar` or `graph hbox`—and the other axis is considered to be categorical. The idea of a categorical axis in this column is not so formal. Indeed, I will emphasize that it is easy to plot categories on a standard y or x axis using `graph twoway`. You will need to do that if the commands just mentioned do not do what you want. There are also some advantages in doing so.

Additionally, graphs may resemble tables if they contain much by way of numbers or text, rather than standard graphic ingredients such as point symbols, line segments, or shaded areas. The stem-and-leaf plots named and popularized by Tukey (1977) and recently discussed in this column by Cox (2007) are a good example. The displays used in meta-analytic studies are another; see Harris et al. (2008) for an excellent Stata-based discussion. Hybridizing graphs and tables in this way remains a fruitful source of innovations in reporting. It has often been suggested (e.g., Cox [2003]) that the distinction between tables and figures owes most to a division of labor imposed by the invention of printing. Different tasks—typesetting and preparing illustrations—became the responsibility of different professions. Modern computing makes that distinction obsolescent, if not obsolete, and creates scope for rethinking forms of reporting.

2   A first example

The Economist (2008) gave a table of data on uses of mobile phones (cell phones, if you prefer) and personal digital assistants (PDAs) by age group (see table 1).

Table 1: Use of mobile phone or PDA to do the following, percent in 2007

	18–29	30–49	50–64	65+
Send or receive text messages	85	65	38	11
Take a picture	82	64	42	22
Play a game	47	29	13	6
Play music	38	16	5	2
Record a video	34	19	8	3
Access the Internet	31	22	10	6
Send or receive email	28	21	12	6
Send or receive instant messages	26	18	11	7
Watch a video	19	11	4	2
At least one of these activities	96	85	63	36

Source: The Economist 2008. Nomads at last: A special report on mobile telecoms. April 12, 4.

Their source: Pew Research Center.

There is no indication in *The Economist* of sample size or method or of where interviews took place. Our focus is entirely on possible presentations of these data. You may consider that a table like this one is perfectly appropriate and adequate for such a small set of numbers with such simple structure: note in particular how uses of all kinds decrease with the age group. Despite that, we can use the data as an example to discuss alternatives.

3   Data entry

The first question is how best to get the data into Stata. The structure may seem obvious from a glance at the table: one string variable and four numeric variables. However, there are other possibilities.

Suppose you typed the various activities on the left of the table into a string variable. When asked for a display, whether table or graph, using that variable Stata would impose its own idea of tidiness and sort those values alphabetically—or, more precisely, alphanumerically. Here, as almost everywhere else, that would not help at all. Alphabetical order is only essential when the sole purpose of a table or graph is looking up individual values and it is irrelevant or confusing otherwise.

Inspection of the table shows a sensible ordering from more popular activities to less popular activities, except for a kind of total at the end. Such ordering by magnitude and appending a total or a ragbag category are common and commendable table habits (see Ehrenberg [1975] for more good advice on tables).

A good Stata answer to this difficulty is to let the activities be value labels to a numeric variable that has the order you desire. (Start with the online help for `label` if value labels are new territory for you.) I am going to add a small twist to this standard advice. You can enter the activities as a string variable and the numeric order as a numeric variable. In the next section, we will see a new command to link the two.

The numeric data also pose a choice of data structure. The table may suggest using four numeric variables, but there can be advantages in having one numeric variable and indicating the age groups by another variable, a long rather than a wide data structure. Whichever way you do it, there is a good chance that some later purpose will make a `reshape` or a `separate` a good idea; see [D] **reshape** or [D] **separate**.

Here it is fortuitous, but fortunate, that the strings "18–29", "30–49", "50–64", and "65+" would always sort the way we want. That would not be true if we had an extra category such as "3–17". A more general way to do it, as before, is to define a numeric variable specifying desired display order and to put the category labels in a string variable.

Typing even a small dataset like this into Stata is a little tedious and error-prone. For other problems, you may be able to read a table in directly to Stata. Or the dataset for a display may result from a reduction command such as `collapse`, `contract`, or `statsby`. For more information, see the corresponding entries in [D]. Otherwise, for data entry, it is worth mastering three Stata tricks that can cut down on the busy work.

First, recall that a variable defining integers 1 and up is obtained by using the observation numbers:

```
. generate seq = _n
```

Second, a more general tool is `egen, seq()` ([D] **egen**). If you followed the advice above of a long data structure you need either 10 blocks of 4 observations or 4 blocks

of 10 observations. Which you choose is immaterial. We will use the first. If you have
not typed anything else, you would need to declare the size of the dataset:

```
. set obs 40
```

Then you could follow with

```
. egen ageorder = seq(), to(4)
. egen activityorder = seq(), block(10)
```

The `egen` function `seq()` is a little unusual. It takes no arguments, but we still need to
spell out that it is a function by giving parentheses `()`. `seq()` creates integer sequences
that by default start at 1 and increase. With this default the first value will be 1, the
second 2, and so on. If an upper limit is specified in the `to()` option and the end of
the dataset has not been reached, the sequence will restart. If a block size is specified,
then values will be repeated in blocks. Thus `ageorder` will start 1 2 3 4 1 2 3 4, and
`activityorder` will start 1 1 1 1 1 1 1 1 1 1 2 2 2 2 2 2 2 2 2 2.

Third, if a string or numeric value is repeated in blocks, you need not type in the
same values repeatedly. Enter just the first value in each block and leave subsequent
entries blank, so that Stata leaves missing values in between, either numeric missing `.`
or the empty string `""`. Then fill in the blanks with a single command

```
. replace whatever = whatever[_n - 1] if missing(whatever)
```

The test `if missing(`*whatever*`)` can apply to numeric or string variables. The idea
is to copy the previous value if a value is missing. As `replace`, and for that matter
`generate`, use the current sort order—a point made explicit by Newson (2004)—this
works with blocks of missing values too. The first missing is replaced by the previous
value, which then becomes nonmissing, so that it can be used in turn to overwrite the
next missing value, and so forth.

A few more general words on data and display: If a novel table or graph form
is being produced, life is often much simpler if the data are cut down first to just
those observations and variables needed. Having to worry about data reduction or
manipulation at the same time as working out the syntax may add up to too much like
hard work. In some cases, it can even be easier to type text and numbers directly into
a completely new dataset than to work out how to get them into new variables within
the existing dataset. Naturally, if you find yourself doing such typing repeatedly, there
is then much more incentive to work out how to automate the tabulation or graphics.

4   Introducing labmask

You can always define value labels explicitly and then link them to a numeric variable, so
long as that numeric variable is integer-valued. However, if you already have a numeric
variable *numvar* and a string variable `strvar` and you want the values of `strvar` to be
the value labels of `numvar`, then another approach is necessary, or at least convenient.

To see this, note that `encode` ([D] **encode**) is of no use here unless you have already defined the labels the way you want. Conversely, if you have defined the labels, you do not need `encode`. The `encode` command is of no use here because its own defensible idea of tidiness is to sort the distinct string values that occur alphanumerically and then link them to integers 1 and up. So you might end up with value labels 1 `"aardvark"`, 2 `"bison"`, 3 `"colugo"`, or whatever. For tables and graphs, that is not usually ideal, as already emphasized.

`labmask` has been invented for this purpose and is published formally with this column. It is more general than the previous problem outline implies, as the values, or if desired the value labels, of one variable can be assigned as the value labels of another. Thus existing string values, value labels, and numeric values can all become the value labels of a numeric variable. The perhaps whimsical name `labmask` arises from the idea that the variable being assigned these value labels acquires a mask, which is what will be seen henceforth. The way `labmask` does that is at root no different from any other way of assigning value labels, but every program author needs to find a program name.

A more formal description of `labmask` is given at the end of this column. For now, just one restriction and one freedom need emphasis. The essential syntax is `labmask` *varname*, `values`(*valuesname*). *valuesname* must not vary within groups defined by the distinct values of *varname* for the observations selected. That should not seem surprising, as it follows from the general idea of a value label. However, there is no rule that the same label may not be assigned to different values of *varname*. In other contexts, that would often seem an odd or even incorrect thing to do, as values and value labels usually correspond one to one. But for graphs, and even for tables, it can be useful. Thus, for example, a year variable might be coded by the party winning at election and those value labels then shown as labels on a graph axis. It is then quite natural that different years may be assigned the same text as a value label.

Assume that somehow we have string variables `activity` and `age` and numeric variables `activityorder`, `ageorder`, and `percent`. In any case, `mobile_uses.dta` released with this column gives the data in this form. Then the string values can be mapped to value labels by

```
. labmask ageorder, values(age)
. labmask activityorder, values(activity)
```

5   Dot charts

For a dataset with similar structure, researchers might consider bar or dot representations. We will focus on the latter, partly because bar graphs do not need much extra publicity. `graph dot` implements the dot charts introduced and discussed by Cleveland (1984, 1993, 1994) and recommended in several guidebooks (e.g., Helsel and Hirsch [1992]; Robbins [2005]; Lang and Secic [2006]). Dot charts should not be confused with the dot plots used to show univariate distributions, as implemented in Stata's `dotplot` command. However, both grow out of the idea of showing numeric values by point symbols on one or more parallel axes. The difference is that programs like `dotplot` allow, or even enforce, binning, and stacking so that distributions can be visualized more easily.

Figure 1 is a graph reached after some trial and error. There remains plenty of scope for small and large changes.

```
. graph dot (asis) percent, over(ageorder) asyvars
> over(activityorder) legend(order(4 3 2 1) row(1))
> marker(1, ms(O) mfcolor(gs1) mlcolor(black))
> marker(2, ms(O) mfcolor(gs5) mlcolor(black))
> marker(3, ms(O) mfcolor(gs9) mlcolor(black))
> marker(4, ms(O) mfcolor(gs13) mlcolor(black))
```

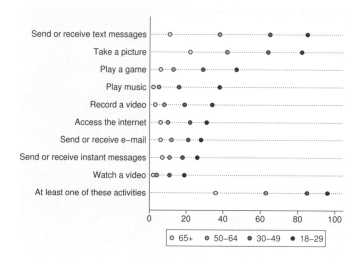

Figure 1: Dot chart of mobile phone and PDA uses, percent 2007. The `asyvars` option forces data for all age groups on to single lines.

Looking at each of the elements of this command in turn:

The data need no further reduction, so we ask for `graph dot (asis) percent`.

`over(ageorder) asyvars` gives us distinctions by `age`, except that we use `ageorder`, which now contains integers 1 and up with values of `age` as value labels, to ensure a logical order. `asyvars` gives different marker symbols: data for the categories of age are treated as if they were separate y variables, meaning responses. (With the extra option `vertical`, the y axis would be truly vertical, but the result is a mess and that path is not pursued here.)

`over(activityorder)` gives us distinctions by `activity`, except that we use `activityorder`, which contains integers 1 and up with values of `activity` as value labels, again to ensure a good order. This option defines a categorical axis, vertical in our case.

The order of options is crucial here. Options `over(activityorder) asyvars over(ageorder)` would define a different graph, although one that seems less helpful, even if tidied up.

`legend(order(4 3 2 1) row(1))` tweaks the legend away from the default. As use declines with age, reversing the order of the labels in the legend seems natural. There is enough space for making the legend a single row, or conversely no need to make it two rows, the default in this case.

The marker options are partly a matter of taste or fine judgment. We are plotting in black and white, which imposes limits that may not constrain your choices for your presentations, especially your talks. As age is an ordered (ordinal) variable, I like to use the same symbol and convey gradations by a choice of gray scales (Cox 2005). Gray scales also have, accidentally but helpfully here, associations with human aging. A black outline for marker symbols ensures that nearly white symbols remain obvious even with a white or nearly white background. Others might find the different colors here insufficient to discriminate between the different ages. Alternatives include varying the marker symbols (e.g., using squares or triangles too) or making the symbols a little larger.

A quite different possibility is to plot on a transformed scale. The most obvious example is a logit scale. For why and how, see Cox (2008).

Let us focus on one detail, the choice of **asyvars**, which forces data points for all age groups onto single lines. If we dispense with that, the syntax is now simpler. With just one kind of marker symbol, there is no need for a legend or specification of different markers. However, the number of data lines increases fourfold, so extra jiggery-pokery is needed to keep the graph legible. One choice is to switch the second `over()` to a `by()` and move to a two column display. See figure 2.

```
. graph dot (asis) percent, over(ageorder)
> by(activityorder, note("") compact cols(2))
```

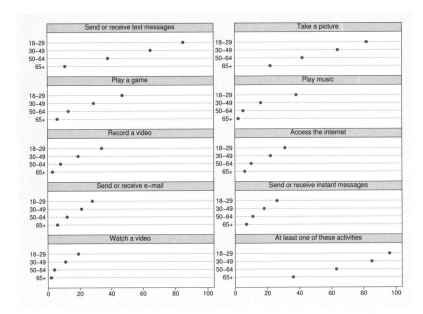

Figure 2: Dot chart of mobile phone and PDA uses, percent 2007. Separate displays for each age group and a switch to by() illustrate some of the trade-offs in graph design.

Another choice is a little more radical. If we force the titles of the by() graphs one ring further out, then a different use of space is shown. The incantations in subtitle() are all important: leaving any out has consequences ranging from the awkward to the grotesque. See figure 3.

```
. graph dot (asis) percent, over(ageorder)
> by(activityorder, col(2) note("") compact)
> subtitle(, pos(9) ring(1) nobexpand bcolor(none) placement(e))
```

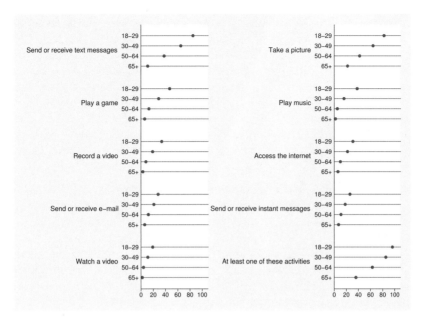

Figure 3: Dot chart of mobile phone and PDA uses, percent 2007. Separate displays for each age group and a switch to `by()`, but now putting graph titles one ring further out.

Yet further tuning clearly remains possible. Some of the activity names could be shortened without much loss, yielding more space for the data region. You may want to experiment. A more crucial detail to carry forward, however, is a realization that you can mimic a categorical axis with suitable use of `by()`. This will be especially useful whenever you need to turn to `twoway` graphs to achieve effects impossible with `graph dot` or its relatives.

6   Why dot charts?

Despite a history about a quarter century long—and no doubt a diligent search would turn up precursors—dot charts have yet to make much of a dent on the popularity of bar charts. I have asked groups of students to choose between bar and dot charts of the same data, and bar charts always win overwhelmingly, but no reason is ever given that does not boil down in one way or another to greater familiarity. My students, like so many others, have been exposed to bar charts from an early age and feel comfortable with them. But dot charts nevertheless have several distinct advantages over bar charts.

Dot charts typically take up less space than bar charts, or at least less ink. The use of space is especially important if you need to show several bars that logically cannot be stacked, as they are not distinct fractions of the same total. Such bars cannot usually be superimposed without some occluding others, so that they must be juxtaposed, taking up more space. In contrast, it is easy to plot several point symbols on the same axis with good use of space and less risk of occlusion.

A bar must have a base, usually zero: clearly, what bar charts show are values encoded by the heights or lengths of bars measured from their bases. A problem arises when values do not vary much, as differences which may be interesting or important are hard to discern. The difference between 102% and 104% may often be a big deal, scientifically or practically. Sometimes a simple change of base is the solution, but otherwise, the need to show a base must be traded off against a temptation to truncate the bars. At worst, readers may struggle to interpret the resulting bar charts correctly, or designers are suspected or accused of misleading graphics. The whole issue is avoided completely by plotting with point symbols. The base of the scale may then even be omitted.

The issue is well illustrated by temperature scales. Weather and climate are often discussed in terms of either Fahrenheit temperatures (in the United States and a few other countries) or Celsius temperatures (everywhere else). Both are interval scales, not ratio scales, meaning that their zero points are arbitrary and ratios make no sense. Even with Celsius, in which zero degrees at least has a physical meaning as the freezing point of water, bar charts are a dubious choice, whether the observed temperatures include zero or negative values, as any choice of base arbitrarily privileges a point on the scale. Point symbols on a numeric scale are not subject to such an objection.

Some readers may be wondering about pie charts. Pie charts are not an option for our example data, as the categories are not mutually exclusive—unless the prospect of 40 pies is appealing. I do not want to add to the criticisms of pie charts by (for example) Tufte (2001) and Cleveland (1994), except to refer to a very well-rounded history and some spirited defense of the form by Spence (2005).

7   Moving to twoway

Despite rafts of options, `graph dot` and its relatives, like any other graph commands, boast only a limited repertoire. There will come a time when you want to do something similar, but nevertheless beyond what is possible with those commands. You will then usually find that moving to `twoway` is the way to go.

First, see that we can do a good job at mimicking `graph dot` with `twoway dot`. See figure 4.

```
. twoway dot percent ageorder,  horizontal
> by(activityorder, col(2) note("") compact)
> subtitle(, pos(9) ring(1) nobexpand bcolor(none) placement(e))
> ysc(reverse) yla(, tlength(0) ang(h) valuelabels) xla(0(20)100)
> xtitle("") ytitle("")
```

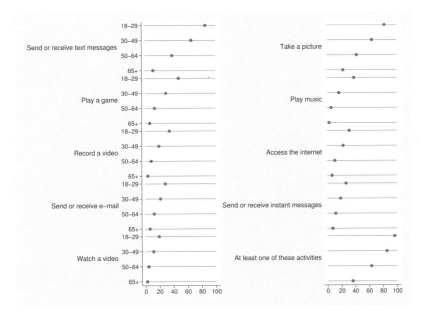

Figure 4: Dot chart of mobile phone and PDA uses, percent 2007. Separate displays for each age group and a switch to by(), but now putting graph titles one ring further out. This was obtained with twoway dot, not graph dot.

We have to work a little to get something acceptable, by insisting on particular choices for axis labels and scales, but no new programming is needed. Axis ticks are natural for numeric scales, but for categorical scales they are too busy, if not misleading, so we should remove them. One subtle, indeed arcane, detail is that with by() in operation using tlength(0) rather than noticks is the way to suppress axis ticks.

The official online help says: "twoway dot is of little, if any use. We cannot think of a use for it, but perhaps someday, somewhere, someone will. We have nothing against the dot plot used with categorical data—see [G-2] **graph dot** for a useful command—but using the dot plot in a twoway context would be bizarre. It is nonetheless included for logical completeness."

Just being able to mimic graph dot is not especially useful and is not enough to overturn such skepticism. But being able to do something else with twoway dot would be more interesting.

One idea from looking at our data table is to notice very simply that our data come as rounded percentages. (If they did not, we could always round them ourselves. Even if the percentages were all very small we could consider a change of scale to, say, per thousand or per million.) This raises the possibility of letting the digits of each value serve as marker labels. All that is needed is some standard twoway technique. See figure 5.

```
. twoway dot percent ageorder, horizontal
> by(activityorder, col(2) note("") compact)
> subtitle(, pos(9) ring(1) nobexpand bcolor(none) placement(e))
> ysc(reverse) yla(, tlength(0) ang(h) valuelabels) xsc(r(0 100)) xla(none)
> xtitle("") ytitle("")
> ms(none) mla(percent) mlabpos(0) mlabsize(*1.2) dcolor(bg)
```

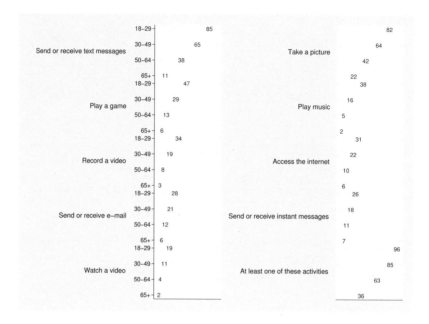

Figure 5: Dot chart of mobile phone and PDA uses, percent 2007. Numeric values as marker labels replace point symbols. This was obtained with `twoway dot`, not `graph dot`.

We suppress the marker symbols and use the numeric values of `percent` as marker labels in their place, in the manner used to produce stem-and-leaf plots and variants (Cox 2007). The *x*-axis labels are now redundant, but we still impose an axis range from 0 to 100 as context. The dots that come with `twoway dot` are a little obtrusive: one way to suppress them is to draw them in background color.

As always, further experiment remains possible. Two paths among others are indicated now. One is to see if `col(1)` yields good results, possibly together with a change in graph size. The other is to note that the end of all our exploring with this example will be to arrive back at what will be a very familiar place to most readers, namely, `scatter`. We can just revise, and indeed simplify, this last graph command to be a `scatter` command. The resulting graph is not shown here, as it is identical to figure 5.

```
. scatter ageorder percent,
> by(activityorder, col(2) note("") compact)
> subtitle(, pos(9) ring(1) nobexpand bcolor(none) placement(e))
> ysc(reverse) yla(, tlength(0) ang(h) valuelabels) xsc(r(0 100)) xla(none)
> xtitle("") ytitle("") ms(none) mla(percent) mlabpos(0) mlabsize(*1.2)
```

You may think the online help remains correct after all.

8 A second example

The following data (table 2) quantify home access to the Internet, again as a percentage, for different groups of people in Sweden in 2002.

Table 2: Home access to the Internet, as a percentage, for different groups of people in Sweden in 2002

Men	66.7
Women	60.3
16–24	75.5
25–34	75.0
35–44	80.0
45–54	75.4
55–64	59.9
65–74	29.8
75–84	10.3
Labourers	49.9
Lower white collar	60.8
Managers and officials	83.5
Entrepreneurs	66.3
Farmers	26.8
Old-age pensioners	22.1

Source: Statistiska centralbyrån. 2003.
Sweden in figures/Sverige i siffror 2004. 52.

The main new twist in this example is that we have three separate little tables. The first, in particular, might seem too trivial to deserve graphing. Many researchers might feel the same way about the other two. But put the tables together and we have the potential for a useful and informative display.

We need a simple way to get what we envisage as a simple graphical display, the only slightly unusual detail being the gaps demarcating the division into separate tables. From the discussion so far, a strategy should be evident:

1. Define a variable containing integers to give a categorical axis. The integers will define axis positions.

2. Define a string variable containing category labels, "Men" and so forth.

3. Use `labmask` to link the two.

4. Define a variable containing the percentages.

5. Use `twoway` because it offers the most flexibility.

6. Underline a principle tacit so far: The percentages are a response or outcome variable, conventionally suggesting their display on the y axis. But what guides axis choice here is a more awkward fact. Once you have more than a few category labels, or have some labels that are rather long, placing categories on the x axis becomes problematic. Either you run out of space, or you are pushed into devices such as reducing text size, alternating label positions, or placing text at an angle to the axis. Even labels such as "Lower white collar" and "Managers and officials", not especially long by many standards, would pose difficulties. Horizontal display of categorical labels is, in this example and indeed in most others, easily the more congenial and convenient choice.

We can use a new tool to make task 1 a little easier, a new command `seqvar` introduced in the next section.

9   Introducing seqvar

`seqvar` is for assigning integer *numlists* to variables. A *numlist* in Stata is a list of numbers, except that extra notation allows concise specification. Thus 1/4 and 10(10)50 are *numlist*s, the first specifying the integers 1 to 4 and the second specifying the integers 10 20 30 40 50. See the online help for `numlist` for more details.

A more formal description of `seqvar` is given at the end of this column. The key idea is that `seqvar` unpacks a supplied *numlist* and assigns it to a variable *varname*, the first value in the *numlist* becoming *varname*[1], the second *varname*[2], and so on.

`seqvar` is designed for one main purpose, under discussion here: easy definition of numeric axis positions as part of setting up what is, in essence, a categorical axis for a graph to be shown using `graph twoway`. That intent does not rule out other uses, but it is likely that other purposes will be better served by use of `egen` functions or direct use of _n and/or _N.

The motivation for restriction to integers may also be apparent. We want to be able to assign value labels to the values produced. While other purposes may require nonintegers, their calculation often entails small precision puzzles arising from the fact that computers work in binary, not decimal (Cox 2006; Gould 2006). Here the programmer wishes to protect himself from a predictable class of problem reports.

The use of `seqvar` may create as many problems as it solves if your data are not in exactly the sort order you need, or if your data include observations that are not to be included in a graph.

10   Exploiting twoway once more

The data for the second example can be found in the file swedish_internet.dta. We want two gaps on the y axis and so we use seqvar to specify a *numlist* with such gaps as the values of a new variable axis. We then use labmask to assign the values of the string variable text as the value labels of axis.

```
. seqvar axis = 1/2 4/10 12/17
. labmask axis, values(text)
```

The first graph with this data is like a dot chart, but is produced using scatter. See figure 6.

```
. scatter axis access,
> yla(1/2 4/10 12/17, valuelabels ang(h) noticks grid glc(gs12))
> ysc(reverse) plotregion(style(outline))
```

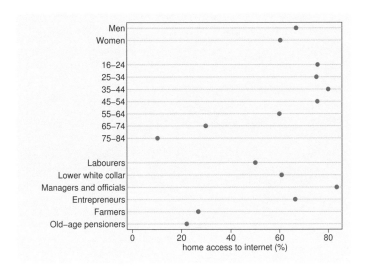

Figure 6: Dot chart showing access to the Internet for various groups of people in Sweden. This was made by scatter, not graph dot.

As the axis is categorical, a reversed scale and omission of ticks are natural. Horizontal axis labels are essential for legibility. I added a grid and an outline to the plot region: those are matters of taste.

If we want finer control of gap sizes, that means using larger integers. The axis positions above have gaps between groups of 1, which is equal to gaps within groups of 1. Suppose you wanted the gaps between groups to be 50% larger than the gaps within groups. Then specifying 10 20 35(10)95 110(10)160 is one of several ways to do that. The numerical axis positions are never explicit, as value labels are always shown instead, so only their relative values matter.

You may find yourself changing the values of a variable like `axis` in this way as you fine-tune the layout of the graph or make small mistakes in working out the sequence. If so, you will need to repeat any application of `labmask`. This is easily done by just reissuing the previous `labmask` command and is still easier than typing out a new `label define` command. Exactly the same applies to any changes of the category labels, such as if you decide to shorten some long labels to give more space to the data region.

Once it is seen how to use `twoway` to display data with this structure, all kinds of variation in graph form become possible using one or more members of the `twoway` family. I restrain myself to three examples.

Horizontal drop lines are used in figure 7. The y-axis grid is suppressed and `recast(dropline) horizontal` added.

```
. scatter access axis, yla(1/2 4/10 12/17, valuelabels ang(h) noticks nogrid)
> ysc(reverse) plotregion(style(outline)) recast(dropline) horizontal
```

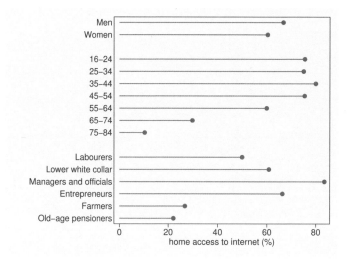

Figure 7: Horizontal drop lines are used to convey magnitude.

Horizontal bars are used in figure 8. Again, the y-axis grid is suppressed, and `recast(bar) horizontal` is added. I prefer not to have bars touching whenever the scale is categorical, so I specified `barw(0.6)`.

```
. scatter access axis, yla(1/2 4/10 12/17, valuelabels ang(h) noticks nogrid)
> ysc(reverse) plotregion(style(outline)) recast(bar) barw(0.6) horizontal
```

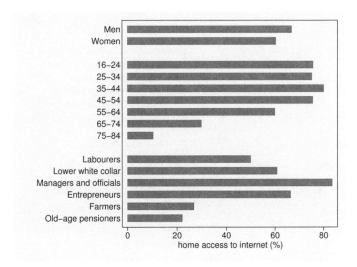

Figure 8: Horizontal bars are used to convey magnitude.

Labels indicating values are added at the ends of bars in figure 9. That requires an overlay of another scatterplot with marker labels and no visible marker symbols. The default position of the marker labels is to the right of the marker symbols that are not visible: that position corresponds to just beyond the top of each bar. Other positions would be equally easy to arrange. For example, labels at the base of each bar would be obtained by generating a variable that was identically 0 and using that for the horizontal coordinate in the second `scatter` command. Given the marker labels, x-axis labels are redundant but we still impose an axis range from 0 to 100 as context.

```
. scatter access axis, yla(1/2 4/10 12/17, valuelabels ang(h) noticks nogrid)
> ysc(reverse) plotregion(style(outline)) recast(bar) barw(0.6) base(0) horizontal
> || scatter axis access, ysc(reverse) ms(none) mla(access)
> legend(off) xla(none) xsc(r(0 100) titlegap(*6))
```

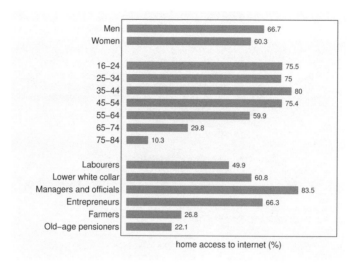

Figure 9: Text labels are added to show magnitude numerically, so that the axis labels can be suppressed.

Finally, `graph dot` can also produce good graphs for this problem. Set up a grouping variable:

```
. generate group = cond(_n < 3, 1, cond(_n < 10, 2, 3))
```

Then use it in an `over()` option. Here we suppress the labels 1 2 3 that would be shown by default. Applying `relabel()` with explicit empty labels will not do that, but `labsize(zero)` does the trick, echoing the `tlength(0)` device used earlier. An alternative would be to add annotations such as "Sex" and "Age group", but they do not seem necessary. The result is similar enough to figure 6 not to be shown here.

```
. graph dot (asis) access, over(axis) over(group, label(labsize(zero))) nofill
```

11   Details on new commands

11.1   labmask

Syntax

labmask *varname* $\left[\,if\,\right]$ $\left[\,in\,\right]$, <u>val</u>ues(*valuesname*) $\left[\,\underline{lbl}name(\textit{lblname})\ \text{decode}\,\right]$

Description

`labmask` assigns the values (or optionally the value labels) of one variable *valuesname* as the value labels of another variable *varname*. Any existing value labels will be overwritten. The idea behind the program name is that henceforth the face that *varname* presents will not be its own, but a mask borrowed from *valuesname*. Thus, for example, a year variable might be coded by party winning at election and those value labels then shown as labels on a graph axis.

varname must take on integer values for the observations selected. *valuesname* must not vary within groups defined by the distinct values of *varname* for the observations selected. However, there is no rule that states that the same label may not be assigned to different values of *varname*.

Options

values(*valuesname*) specifies a variable whose values (by default) or value labels (optionally) will be used as the value labels of *varname*. values() is required.

lblname(*lblname*) specifies that the value labels to be defined will have the label name *lblname*. The default is that they will have the same name as *varname*. Any existing value labels for the same values will be overwritten in either case.

decode specifies that the value labels of *valuesname* should be used as the value labels of *varname*. The default is to use the values of *valuesname*.

11.2 seqvar

Syntax

seqvar *varname* = *numlist* [, replace]

Description

seqvar assigns the values in an integer *numlist* to the successive observations (observation number 1 and up) of a variable *varname*. If the *numlist* contains fewer values than the number of observations in the dataset, the remaining values will be defined as missing. If the *numlist* contains more values than the number of observations, only as many values as can be assigned will be used and the rest will be ignored. The *numlist* must include integers only and may not include missing values.

Option

replace specifies that if *varname* exists then all its values will be replaced. This option is essential for overwriting an existing variable. Nonmissing values will be replaced by missing values whenever the *numlist* does not specify sufficient integers to be assigned.

12   Conclusion

Table-like graphs can be interesting, useful, and even mildly innovative. This column has outlined some Stata techniques for producing such graphs.

If one or more variables, often controlling or contextual, are categorical, then `graph box`, `graph bar`, `graph pie`, and `graph dot` are possible commands. Of these commands, `graph dot` is likely to be the most under-appreciated.

Using `by()` with various choices is a good way to mimic a categorical axis. That applies to many `graph` commands.

When the `bar` or `dot` commands are not flexible enough to do what you want, moving to `twoway` is usually the answer. `twoway` is enormously flexible.

`labmask` is a new tool for the situation in which you want the values, or the value labels, of one variable to be the value labels of another variable. It thus automates the definition and attachment of a set of value labels.

`seqvar` is a new tool for assigning integers to a variable. Used with `labmask`, it allows you to produce categorical axes with gaps to indicate groupings.

The underlying theme is hardly exhausted by this column. Plots of confidence intervals or other intervals indicating uncertainty often benefit from similar devices, but that is a large enough story to merit separate treatment. Overlaying of two or more graphs looms much larger in that story than it does in this one.

13   Acknowledgments

I have benefited from questions, suggestions, and interest in this territory from Kit Baum, Ronán Conroy, Friedrich Huebler, Hiroshi Maeda, and Austin Nichols. Vince Wiggins told me about the `tlength(0)` trick. Omitting ticks by setting their length to 0 is all too obvious in retrospect.

14   References

Cleveland, W. S. 1984. Graphical methods for data presentation: Full scalebreaks, dot charts, and multibased logging. *American Statistician* 38: 270–280.

―――. 1993. *Visualizing Data*. Summit, NJ: Hobart.

―――. 1994. *The Elements of Graphing Data*. Rev. ed. Summit, NJ: Hobart.

Cox, N. J. 2003. Speaking Stata: Problems with tables, Part I. *Stata Journal* 3: 309–324.

―――. 2005. Stata tip 27: Classifying data points on scatter plots. *Stata Journal* 5: 604–606.

————. 2006. Stata tip 33: Sweet sixteen: Hexadecimal formats and precision problems. *Stata Journal* 6: 282–283.

————. 2007. Speaking Stata: Turning over a new leaf. *Stata Journal* 7: 413–433.

————. 2008. Stata tip 59: Plotting on any transformed scale. *Stata Journal* 8: 142–145.

Ehrenberg, A. S. C. 1975. *Data Reduction: Analysing and Interpreting Statistical Data*. London: Wiley.

Gelman, A., C. Pasarica, and R. Dodhia. 2002. Let's practice what we preach: Turning tables into graphs. *American Statistician* 56: 121–130.

Gould, W. 2006. Mata Matters: Precision. *Stata Journal* 6: 550–560.

Harris, R. J., M. J. Bradburn, J. J. Deeks, R. M. Harbord, D. G. Altman, and J. A. C. Sterne. 2008. metan: Fixed- and random-effects meta-analysis. *Stata Journal* 8: 3–28.

Helsel, D. R., and R. M. Hirsch. 1992. *Statistical Methods in Water Resources*. Amsterdam: Elsevier. http://pubs.usgs.gov/twri/twri4a3/.

Lang, T., and M. Secic. 2006. *How to Report Statistics in Medicine: Annotated Guidelines for Authors, Editors, and Reviewers*. Philadelphia: American College of Physicians.

Newson, R. 2004. Stata tip 13: generate and replace use the current sort order. *Stata Journal* 4: 484–485.

Robbins, N. M. 2005. *Creating More Effective Graphs*. Hoboken, NJ: Wiley.

Spence, I. 2005. No humble pie: The origins and usage of a statistical chart. *Journal of Educational and Behavioral Statistics* 30: 353–368.

Statistiska centralbyrån. 2003. *Sweden in Figures/Sverige i Siffror 2004*. Örebro: Statistiska centralbyrån.

The Economist. 2008. Nomads at last: A special report on mobile telecoms. April 12, 4.

Tufte, E. R. 2001. *The Visual Display of Quantitative Information*. 2nd ed. Cheshire, CT: Graphics Press.

Tukey, J. W. 1977. *Exploratory Data Analysis*. Reading, MA: Addison–Wesley.

About the author

Nicholas Cox is a statistically minded geographer at Durham University. He contributes talks, postings, FAQs, and programs to the Stata user community. He has also coauthored 15 commands in official Stata. He wrote several inserts in the *Stata Technical Bulletin* and is an editor of the *Stata Journal*.

The Stata Journal (2009)
9, Number 3, pp. 478–496

Speaking Stata: Creating and varying box plots

Nicholas J. Cox
Department of Geography
Durham University
Durham City, UK
n.j.cox@durham.ac.uk

Abstract. Box plots have been a standard statistical graph since John W. Tukey and his colleagues and students publicized them energetically in the 1970s. In Stata, `graph box` and `graph hbox` are commands available to draw box plots, but sometimes neither is sufficiently flexible for drawing some variations on standard box plot designs. This column explains how to use `egen` to calculate the statistical ingredients needed for box plots and `twoway` to re-create the plots themselves. That then allows variations such as adding means, connecting medians, or showing all data points beyond certain quantiles.

Keywords: gr0039, box plots, dispersion diagrams, distributions, egen, graphics, percentile, quantile, range bars, twoway

1 Box plots

1.1 Origins

Box plots were so named by John W. Tukey and were publicized energetically within statistics by him, his colleagues, and his students from the 1970s on (e.g., Tukey [1972, 1977]; Velleman and Hoaglin [1981]; and Hoaglin, Mosteller, and Tukey [1983]). Box plots spread beyond statistics into several quantitative sciences through their own literature (e.g., Kleiner and Graedel [1980] and Cox and Jones [1981]). The publicity was so successful that the box plot is now widely regarded as a standard statistical graph. It appears in most introductory statistical texts; indeed, the exceptions to this rule (e.g., Freedman, Pisani, and Purves [2007]) are more striking than the examples. Further, the box plot is often assumed not to need explanation beyond such texts.

Box plots had several under-appreciated precursors under different names, including range bars (Spear 1952, 1969) and dispersion diagrams in geography and climatology (e.g., Crowe [1933] and Monkhouse and Wilkinson [1971]). Despite this earlier history, my guess is that box plots would not now be nearly so popular without Tukey's reinvention and propaganda.

1.2 Purpose

Stata users wishing to see box plots can call upon `graph box` or `graph hbox`. The manual entry [G-2] **graph box** explains several ways of tuning that command. Mitchell (2008) gives many examples of possible results and the code to get them. This column focuses on showing what to do whenever you want some variation on the standard design that cannot be met with `graph box` or `graph hbox`. To show that, we must understand how to re-create box plots using `graph twoway`. It is very much a case of *reculer pour mieux sauter*.

1.3 Structure

Let us first remind ourselves of the structure of a box plot by using the life expectancy data shipped with Stata. We will compare life expectancy in 1998 for three groups of countries: in Europe and Central Asia, North America, and South America (figure 1). We use `graph box`. Here and subsequently we will spell out a preference for horizontal axis labels.

```
. sysuse lifeexp
. label variable lexp "Life expectancy (years)"
. graph box lexp, over(region) yla(, ang(h))
```

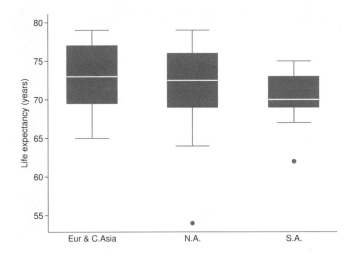

Figure 1: Box plots of life expectancy in 1998 for various countries in three regions

The main ingredient of a box plot is the eponymous box, used to indicate the lower and upper quartiles of the variable or group being plotted against a magnitude scale. The median is represented by a line subdividing the box, or, alternatively, by a point symbol. The length of the box thus represents the interquartile range (IQR). Tukey used a variety of alternative terms for both the quartiles (hinges, fourths, etc.) and their difference, the range or spread between them, but most such terms were adopted only locally or briefly and have long since faded away. It seems simpler now to revert to the classical terms of quartiles and IQR. Whatever the terminology, recall that numerous slightly different calculation rules exist for quartiles and quantiles or percentiles generally (Frigge, Hoaglin, and Iglewicz 1989; Hyndman and Fan 1996). The different rules explain some of the differences in box plots from different software, but otherwise are not of great interest. Stata's rule is set out in [R] **summarize**. Among other details, note that any practical rule must extend to data with weights assigned.

Box plots differ in what else may be shown outside the box. `graph box` and `graph hbox` by default follow what is perhaps the most common recipe (Tukey 1977):

1. Lines, often called whiskers, are drawn to span all data points within 1.5 IQR of the nearer quartile. That is, one whisker extends to include all data points within 1.5 IQR of the upper quartile and stops at the largest such value, while the other whisker extends to include all data within 1.5 IQR of the lower quartile and stops at the smallest such value. Tukey called the outer limits of the whiskers *adjacent values*. The whiskers also explain his alternative term, *box-and-whiskers plots*. Note that either whisker could be of zero length. In practice, that will occur only with very small datasets or heavily tied data.

2. Any data points beyond the whiskers are shown individually and often labeled informatively.

De Veaux, Velleman, and Bock (2008, 81) record Tukey's laconic reply when asked the reason for 1.5: 1 would be too small and 2 would be too large. Evidently, the choice of multiplier gives an informal but objective rule for outlier identification. Any choice is a compromise between revealing too much (flagging data points that are of neither statistical nor scientific concern) and revealing too little (missing data points that require thought or action). Dümbgen and Riedwyl (2007) recently discovered a clever way of justifying 1.5, but experience that it often works quite well is a more compelling basis for the rule.

This kind of box plot, and indeed most other kinds, thus conveys information about level (median); spread (interquartile range and range are both represented directly); symmetry or asymmetry about the median both within and beyond the central half of the data; and, on its own definition, possible outliers. It is thus a fairly information-rich graphical reduction of key quantiles (or of order statistics, if you prefer).

That may be the most common recipe, but many others have been entertained. McGill, Tukey, and Larsen (1978) suggested two refinements: varying the width of boxes to indicate group sizes and notching boxes to indicate approximate confidence intervals. Harris (1999, 57) even reported that some box plots are based at least in part on mean and standard deviation. It is natural to hope that different conventions are all explained clearly for the benefit of readers, but unfortunately, that is often not the case. For example, several authors in the collection edited by Chen, Härdle, and Unwin (2008) use differing varieties of box plots, but the differences are typically unexplained.

However, many variations encountered appear to be essentially cosmetic. In particular, box plots may be horizontal as well as vertical. There can be a small struggle between the convention of showing response or outcome variables increasing vertically and the desire that text labels explaining variables or groups can be spelled out fully and legibly. Whatever the reasoning, Stata users can reach for `graph hbox` if they prefer horizontal alignment. As a matter of careful and conscious design, the change between typing `box` and `hbox` is the only change that need be made. Contrary to mathematical custom, the y axis of box plots in Stata is considered to be whichever axis the response is plotted against. (`graph bar` and `graph hbar` are related in exactly the same way.)

1.4 Utility

Box plots can be very useful, particularly for comparison, especially if the number of variables or groups is nearer 20 or 200 rather than 2. But if you have just a few variables or groups, you have enough space for the greater detail of (say) histograms, dot plots, density traces, or quantile or distribution plots. And because they are reductions of the data, box plots may be uninformative about key details. They tend to perform poorly whenever data are highly skewed—which in many fields is overwhelmingly usual. Naturally, one simple answer to skewness is to transform data. If box plots of a variable are highly asymmetric, then roots or logs or reciprocals are likely to improve matters considerably.

There are deeper problems yet. What is so special about quartiles, in particular? Medians have a clear statistical role as defining midpoints on distribution functions, and they are natural and resistant summaries for (approximately) symmetric distributions. Quartiles take the median idea one step further by being medians of each half of the distribution, but beyond that, their role is much less evident. Simplicity of definition and familiarity from early teaching do not add up to a statistically natural role. In any case, if half the data lie inside the boxes, then half too lie outside the boxes, yet that half—often statistically or scientifically the more important half—is represented in a mostly generalized way within box plots.

So, other quantiles besides quartiles may well be as or more worthy of display. That argument leads ultimately to displaying all quantiles, a tactic discussed in other issues of the *Stata Journal* (Cox 2005, 2007).

With a nod of gratitude for an example given by Wainer (1990, 345), figure 2 points out one further weakness of box plots.

Figure 2: An innocent-looking box plot with a surprise wrapped inside

Asked what can be inferred about the distribution from this plot, even very experienced data analysts typically mutter something about a short-tailed symmetric unimodal distribution. But the box plot clearly implies that the average density in the tails is much greater than that in the middle, so the best inference should be something like a U-shaped distribution. My guess is that although respondents are all familiar with the main idea of box plots, they are being misled by the subdued representation of the tails. Guessing apart, no detailed histogram, density trace, or quantile plot would be guilty of such ambiguity. More generally, box plots inevitably gloss over bimodality or multimodality or granularity of distribution.

To reveal the small surprise, figure 2 is based on a set of quantiles from a beta distribution:

```
. generate y = invibeta(0.6, 0.6, (_n - 0.5) / _N)
```

With these parameter values, the distribution is indeed U-shaped, as the histogram in figure 3 shows more clearly.

```
. histogram y, width(0.1) start(0) horizontal yla(, ang(h))
```

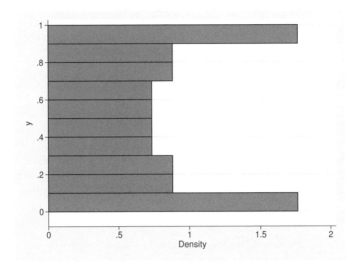

Figure 3: The distribution underlying the innocent-looking box plot: a U-shaped beta distribution

2 Using twoway to create box plots

2.1 Ingredients

To re-create a box plot from scratch given some data, we need to calculate the basic summary statistics. Here the **egen** command is your friend, particularly because its **by()** option allows recording of results for two or more groups. The **by()** option is undocumented in favor of doing things with **by** *varlist*:, but it is supported for those **egen** functions of concern to us here. Either way, using **by:** as prefix is exactly equivalent. See the online help or [D] **egen** for more details on that command. A tutorial discussing **egen** is available in Cox (2002).

The median and quartiles are easiest:

```
. egen median = median(lexp), by(region)
. egen upq = pctile(lexp), p(75) by(region)
. egen loq = pctile(lexp), p(25) by(region)
```

We could now get the IQR by subtraction, **upq - loq**, which would be more efficient, but we will mention that it has its own **egen** function.

```
. egen iqr = iqr(lexp), by(region)
```

In fact, we do not strictly need the IQR, as will become clear shortly, but if you like box plots, you might as well know ways of getting the IQR easily into a variable.

The upper and lower limits of the whiskers require a little more thought. Here is one way to get them. The upper limit is the largest value not greater than `upq + 1.5 * iqr`. That can be calculated in one line:

```
. egen upper = max(min(lexp, upq + 1.5 * iqr)), by(region)
```

That one line could bear some deconstruction, however. The outer `max()` is an `egen` function, as the context implies. The inner `min()` is emphatically not another `egen` function, as might be guessed: it is just the standard Stata function `min()`. Why is that allowed here? Because the syntax of `egen` allows here an arbitrary expression, indicated in the syntax diagram by *exp*. Often that expression is just one variable name, but it could be more complicated. Here the entire expression is `min(lexp, upq + 1.5 * iqr)`. The expression could have been `min(lexp, upq + 1.5 * (upq - loq))`, showing that the IQR variable is indeed redundant.

As before, the `by(region)` option ensures that maximums for the expression supplied are calculated separately for each region.

For lower limits of whiskers, we can use the same tactic, except for swapping minimum and maximum:

```
. egen lower = min(max(lexp, loq - 1.5 * iqr)), by(region)
```

We now have in hand all the ingredients we need. But one basic point needs emphasis. By construction, the values for the median, quartiles, and upper and lower limits of the whiskers are repeated for each distinct value of `region`. If instead of comparing groups we were comparing variables, then values would be repeated for each observation. Unless we do nothing further, the graphical consequence will be repeated plotting of the same information, which could be time-consuming and which leads to unnecessarily bloated graph files. It would be important to do something about that in any program with pretensions to efficiency, but for our purposes, we will set this detail aside, beyond noting that `collapse` and `egen, tag()` offer some solutions.

2.2 Assembly

Let us jump immediately to a tolerable mock-up of a box plot and then talk through all the details. Let me also stress that even Stata graph experts never write down code just like this, unless they happen to have solved the problem a few minutes earlier and have excellent memory. To get here requires much experiment and consultation of the help. Figure 4 shows the result.

```
. twoway rbar med upq region, pstyle(p1) blc(gs15) bfc(gs8) barw(0.35) ||
> rbar med loq region, pstyle(p1) blc(gs15) bfc(gs8) barw(0.35) ||
> rspike upq upper region, pstyle(p1) ||
> rspike loq lower region, pstyle(p1) ||
> rcap upper upper region, pstyle(p1) msize(*2) ||
> rcap lower lower region, pstyle(p1) msize(*2) ||
> scatter lexp region if !inrange(lexp, lower, upper), ms(Oh) mla(country)
> legend(off)
> xla(1 `" "Europe and" "Central Asia" "´ 2 "North America" 3 "South America",
> noticks) yla(, ang(h)) ytitle(Life expectancy (years)) xtitle("")
```

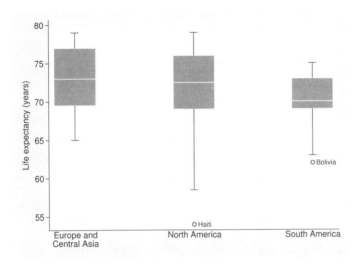

Figure 4: Box plots of life expectancy in 1998 for various countries in three regions, but constructed entirely using `twoway`

Now the commentary:

1. The details may look scary in total, but note first the strategy, which is divide and conquer. Different parts of `twoway` are enlisted to draw different parts of the graph. Similarly, divide and conquer is the strategy to understand the code. There is clearly no need to try to reproduce all the details produced by `graph box` if you prefer something different.

2. `region` is a numeric variable, so we can plot against it. Its values are 1, 2, and 3, and value labels are attached, so it is already in good condition for graphics. If you had a variable that was not in good condition, say, because it was a string variable or a numeric variable that needed tidying up, then creating a new variable with `egen, group()` with its `label` option is the best way to proceed. `encode` is an alternative for string variables.

3. `pstyle(p1)` is a simple trick to enforce general consistency of style. You can then depart from whatever results in your preferred directions.

4. The boxes are drawn with `twoway rbar`, one from the median to the upper quartile and one from the median to the lower quartile. A light outline color, `blcolor(gs15)`, is sufficient to indicate where the medians are. I chose as a matter of personal taste a lighter color for the bar fill than the default in the `sj` scheme. Light color for fill and dark color for outline are equally acceptable statistically, and perhaps preferable aesthetically. `barwidth(0.35)` reflects my personal taste: I regard the default boxes of `graph box` as a little fat. If the values of the categorical variable did not differ by 1, a quite different bar width would be needed.

5. The whiskers are drawn with `twoway rspike`, one from the lower quartile to the lower end of the whiskers, and one from the upper quartile to their upper end.

6. The whiskers are capped using `twoway rcap`. Note that there is no typo in `rcap upper upper region` or `rcap lower lower region`. The code was not `rcap upq upper region` or `rcap loq lower region` to ensure that no caps are visible interfering with the box. The marker size is twice default, but even so is much less than the default of `graph box`, to say nothing of what can be obtained using its `capsize()` option.

7. Clearly, the caps could be omitted if so desired, simply by omitting the calls to `twoway rcap`. Why does the standard box plot design include them? It seems to be an admission of weakness, namely, that the whiskers might be overlooked if the graph did not emphasize where they end.

8. Data points beyond the whiskers are shown using `scatter`. Hollow circles given by `ms(Oh)` are a personal choice as suitably prominent yet tolerating overlap well (think of the overlapping rings of the Olympic symbol). Note the simple logic: points within the range of the boxes and whiskers are `inrange(lexp, lower, upper)` and so points beyond them are the logical complement, obtained by negation, `!`. See Cox (2006) for more on `inrange()` if so desired. Putting this into words as "not in range" is a simple way of underlining what is being done.

9. Such data points are labeled using marker labels, `mla(country)`. In this case, defaults work fine. In other cases, we might want to tune marker label size or other properties, as later examples will make clear.

10. All the different `twoway` calls produce a complicated `legend`, which we just suppress. So many different variables are being portrayed, from `twoway`'s point of view, that we have to add our own y-axis title.

11. In this particular case, the value labels attached to `region` are over-abbreviated, so we step in and provide our own. I agree with the designer of `graph box` that axis ticks serve no useful purpose when distinct categories are being shown. The default `xtitle()` would be the variable name `region`, which also is dispensable here. (In other contexts, I routinely suppress variable names indicating date or year when axis labels such as 1990 or 2000 make abundantly clear what is being shown.)

2.3 Horizontal

Clearly, we need to know how to produce horizontal box plots too. Here is a first stab, with the result in figure 5:

```
. twoway rbar med upq region, horiz pstyle(p1) blc(gs15) bfc(gs8) barw(0.35) ||
> rbar med loq region, horiz pstyle(p1) blc(gs15) bfc(gs8) barw(0.35) ||
> rspike upq upper region, horiz pstyle(p1) ||
> rspike loq lower region, horiz pstyle(p1) ||
> rcap upper upper region, horiz pstyle(p1) msize(*2) ||
> rcap lower lower region, horiz pstyle(p1) msize(*2) ||
> scatter region lexp if !inrange(lexp, lower, upper), mla(country) legend(off)
> yla(1 `" "Europe and" "Central Asia" "´ 2 "North America" 3 "South America",
> ang(h) noticks) xtitle(Life expectancy (years)) ytitle("")
```

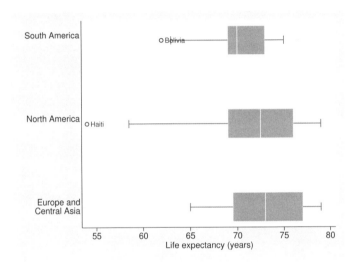

Figure 5: Horizontal box plots of life expectancy in 1998 for various countries in three regions, but constructed entirely using `twoway`

The necessary changes are to add the `horizontal` option to calls to `rbar`, `rspike`, and `rcap` and to swap y and x within the call to `scatter` (variables are swapped, `x` options become `y` options, and vice versa).

The result is most of the way to where we want to be. The marker labels would be better lifted clear of the whiskers. The x axis also needs to be lengthened a little to give enough space for the text `Haiti`. A little experiment shows that the extra options `xsc(r(53,.))`, `mlabpos(12)`, and `mlabgap(1.5)` give those improvements; see figure 6.

```
. twoway rbar med upq region, horiz pstyle(p1) blc(gs15) bfc(gs8) barw(0.35) ||
> rbar med loq region, horiz pstyle(p1) blc(gs15) bfc(gs8) barw(0.35) ||
> rspike upq upper region, horiz pstyle(p1) ||
> rspike loq lower region, horiz pstyle(p1) ||
> rcap upper upper region, horiz pstyle(p1) msize(*2) ||
> rcap lower lower region, horiz pstyle(p1) msize(*2) ||
> scatter region lexp if !inrange(lexp, lower, upper), mla(country)
> mlabpos(12) mlabgap(1.5)  xsc(r(53, .)) legend(off)
> yla(1 `" "Europe and" "Central Asia" "´ 2 "North America" 3 "South America",
> ang(h) noticks) xtitle(Life expectancy (years)) ytitle("") ;
```

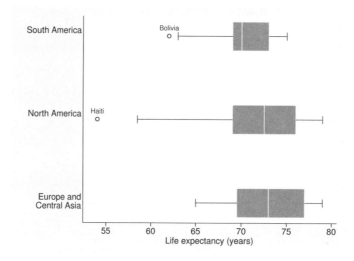

Figure 6: Horizontal box plots of life expectancy in 1998 for various countries in three regions, with improved positioning of marker labels for outliers

3   Moving beyond standard designs

Provided that you are broadly familiar with how twoway works, you should now have a sense that a small new world is open before you, in which you can add to, subtract from, or otherwise vary box plot designs exactly as you wish. If you do this repeatedly, you will want to encapsulate code for favored designs in a do-file or program. Explaining that further would take us beyond the main story, but both the User's Guide and Baum (2009) are excellent sources of advice and examples.

3.1   Adding means

One common request, on Statalist and elsewhere, is to add means to box plots. For this, you need an extra variable containing means. egen is again convenient:

```
. egen mean = mean(lexp), by(region)
```

We need to add a `scatter` call to the code above:

```
scatter region mean, ms(Dh) msize(*2) ||
```

A simple but crucial detail is plotting the means after, and therefore on top of, the boxes. Usually, although not inevitably, means will lie between the quartiles, and so their symbols would disappear under the boxes otherwise. Figure 7 shows the result.

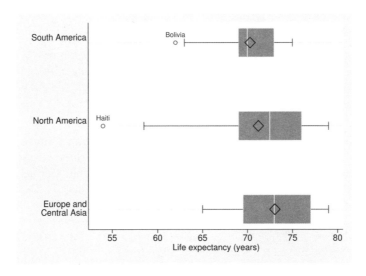

Figure 7: Horizontal box plots of life expectancy in 1998 for various countries in three regions; diamond symbols indicate means

3.2 Connecting medians

Another common request is connecting medians. One context for this could be that the box plots indicate variation within time periods. The connected medians thus would emphasize variation between time periods. This request is met as previously, by adding another `twoway` call such as

```
line median timevar, lw(*2)
```

or

```
line timevar median, lw(*2)
```

depending on whether plots are vertical or horizontal. Emphasis is added if and as desired, here by doubling line width. As before, plot connecting lines after, and so on top of, boxes.

3.3   Unequal spacing

Nothing in the `twoway` route to box plots commits you to equal spacing of box plots. Unequal spacing is perfectly possible: you just specify the positions of the box plots. Binning of responses or residuals in unequal intervals of a covariate is one large class of possible examples.

3.4   Variable width

If there were a desire for boxes of variable width, that could be met by repeated calls to `twoway rbar` with differing `barwidth()` options. `barwidth()` requires a single number as argument, and does not accept a numeric variable indicating width.

3.5   Percentile-based whiskers

Let us now imagine a different design in which whiskers are drawn out to 10% and 90% points. Cleveland (1985) showed such box plots. They have three advantages over the standard design. First, the definition of whiskers is of the same kind as the definition of boxes. Second, almost always, we see some detail in the tails. The exceptions when there is heavy tying in one or the other tail are also discernible. Third, to a very good approximation, drawing such box plots commutes with any monotonic transformation so that, for example, the box plot of a logged variable is the log of the box plot of the variable on the original scale. Some minor inaccuracy may arise in practice because quantiles may be calculated as the average of two order statistics: see the FAQ at http://www.stata.com/support/faqs/graphics/boxandlog.html for more on this thorny little detail.

Evidently, the choice of 10% and 90% is in no sense compulsory: other values may suit some purposes better.

We will also ensure that all points outside the whiskers are labeled. Because we are in complete control, we will go back to vertical, reverse box coloring and drop those whisker caps that we do not much like.

We know how to get further percentiles:

```
. egen p10 = pctile(lexp), p(10) by(region)
. egen p90 = pctile(lexp), p(90) by(region)
```

There are no new tricks needed for the graph, or so we might think; see figure 8.

```
. twoway rbar med upq region, pstyle(p1) bfc(gs15) blc(gs8) barw(0.35) ||
> rbar med loq region, pstyle(p1) bfc(gs15) blc(gs8) barw(0.35) ||
> rspike upq p90 region, pstyle(p1) ||
> rspike loq p10 region, pstyle(p1) ||
> scatter lexp region if !inrange(lexp, p10, p90), ms(Oh) mla(country)
> mlabgap(1.5) legend(off)
> xla(1 `" "Europe and" "Central Asia" "´ 2 "North America" 3 "South America",
> noticks) yla(, ang(h)) ytitle(Life expectancy (years)) xtitle("")
```

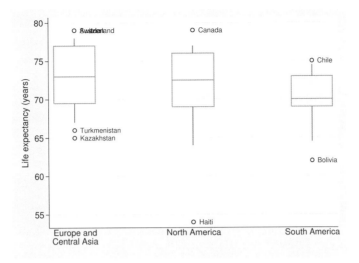

Figure 8: Box plots of life expectancy in 1998 for various countries in three regions; whiskers extend to 10% and 90% points of the distribution

A little mess of marker labels turns out to arise because Austria, Sweden, and Switzerland tie at 79 years. Some experimenting indicates that we can just rotate two of those labels away from the default position. Figure 9 is the improved graph.

```
. generate pos = cond(country == "Austria", 1, cond(country == "Sweden", 4, 3))
. twoway rbar med upq region, pstyle(p1) bfc(gs15) blc(gs8) barw(0.35) ||
> rbar med loq region, pstyle(p1) bfc(gs15) blc(gs8) barw(0.35) ||
> rspike upq p90 region, pstyle(p1) ||
> rspike loq p10 region, pstyle(p1) ||
> scatter lexp region if !inrange(lexp, p10, p90), ms(Oh) mla(country)
> mlabgap(1.5) legend(off) mlabvpos(pos)
> xla(1 `" "Europe and" "Central Asia" "´ 2 "North America" 3 "South America",
> noticks) yla(, ang(h)) ytitle(Life expectancy (years)) xtitle("")
```

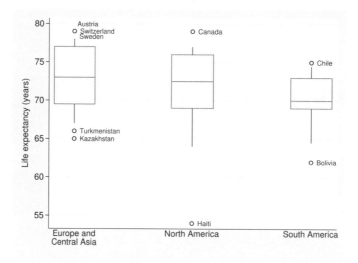

Figure 9: Box plots of life expectancy in 1998 for various countries in three regions; whiskers extend to 10% and 90% points of each distribution; marker labels for Austria and Sweden have been moved to avoid overlap

3.6   Other data structures

So far, we have considered only the case of one response variable, subdivided by groups of a categorical variable. Box plots are often needed for other data structures. We need to see that they are also within reach given a little technique.

Another dataset shipped with Stata contains temperature data for 956 cities in the United States, including variables `tempjan` and `tempjuly` indicating mean monthly temperatures for January and July. The cities are classified coarsely by `region` and more finely by `division`. We will produce box plots of the Cleveland (1985) kind for the two temperature responses `tempjan` and `tempjuly`, subdivided by division and month.

We could just superimpose box plots for `tempjan` and `tempjuly`, but a `reshape` of the data makes matters easier thereafter. `reshape` requires a unique identifier, so we put the observation number into a new variable to act as a pacifier. The identifier will play no role thereafter in our graphics. See the online help or [D] **reshape** if you need more discussion.

```
. sysuse citytemp, clear
. generate id = _n
. reshape long temp, i(id) string j(month)
```

The new string variable `month` takes on two values, `jan` and `july`. The summary statistics come from `egen`. The extra twist that we need to distinguish both `division` and `month` is easily satisfied:

```
. egen median = median(temp), by(division month)
. egen loq = pctile(temp), p(25)  by(division month)
. egen upq = pctile(temp), p(75)  by(division month)
. egen p10 = pctile(temp), p(10)  by(division month)
. egen p90 = pctile(temp), p(90)  by(division month)
```

division is an integer variable with values from 1 to 9 and value labels attached. To show box plots for January and July side by side, we just need a position variable in which months are offset. Cui (2007) gives further discussion of this simple trick. We still want to use the value labels of division, so we assign them to the new variable.

```
. generate division2 = division + cond(month == "jan", -0.2, 0.2)
. label values division2 division
```

The code is now very similar to previous examples. Figure 10 gives the result.

```
. twoway rbar median upq division2, bfc(gs15) blc(gs8) barw(0.35) ||
> rbar median loq division2, bfc(gs15) blc(gs8) barw(0.35) ||
> rspike loq p10 division2 ||
> rspike upq p90 division2 ||
> scatter temp division2 if !inrange(temp, p10, p90), ms(o) legend(off)
> xaxis(1 2) xla(1/9, valuelabel noticks grid axis(1))
> xla(1/9, valuelabel noticks axis(2)) xtitle("", axis(1)) xtitle("", axis(2))
> yaxis(1 2) yla(14(18)86, ang(h) axis(2))
> yla(14 "-10" 32 "0" 50 "10" 68 "20" 86 "30", ang(h) axis(1))
> ytitle(mean temperature ({c 176}F), axis(2))
> ytitle(mean temperature ({c 176}C), axis(1))
> ysc(titlegap(0) axis(1)) ysc(titlegap(0) axis(2))
> plotregion(lstyle(p1))
```

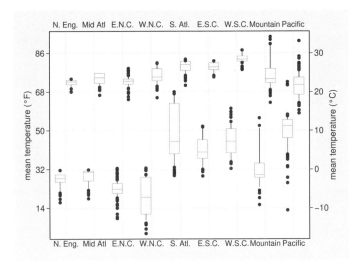

Figure 10: Box plots of mean temperatures in January (left plot in each group) and July (right plot in each group) for various places in divisions of the United States; whiskers extend to 10% and 90% points of each distribution

Here we know that each division will take up less space than in previous graphs, so we bump up the marker symbol to `ms(o)` so that they remain visible.

We also add a few `twoway` flourishes. Horizontal axis labels at the top as well as the bottom of the graph ease look-up of division labels. 0°F is of less importance than 32°F as a reference temperature. We also align equivalent temperatures on Fahrenheit and Celsius scales. Note the trick to get the degree symbol (Cox 2004).

Incidentally, when we look at the box plot to learn something about the data, we see that the upper tail of Mid-Atlantic July temperatures is curiously truncated. Inspection of the data shows that 20 places are all given a mean July temperature of 76.8°F. This is the highest temperature observed for that division but is also the 90% point, because those 20 are more than 10% of the places in the division. Thus no places are plotted as having a higher temperature than the 90% point. Hence the graph is correct in terms of the data, but the graph has also told us something new about the data, which is as it should be. People more familiar with how the U.S. Census reports temperature data may be able to throw more light on this little mystery.

3.7 Convenience and efficiency

The stress in this column has been on getting results conveniently using standard commands. It is nevertheless proper to repeat a note of caution sounded earlier. The commands used here are not the most efficient way to get box plots, nor will the graph files produced be as lean as they could be. For small or moderate datasets, you would have to strain to notice that, but otherwise you might be bitten. Industrial-strength alternatives to these commands would need to work at lower levels to optimize speed and storage, by replacing calls to `egen` with direct calls to `summarize` and by ensuring that the information defining box plot ingredients is not duplicated unnecessarily.

4   Conclusion

`graph box` and `graph hbox` are very useful commands, but they only do what they claim to do. This column has shown an alternative way to create, and then to vary, box plots, using `egen` for calculations and `twoway` for graphics. Once the problem is broken down into components, it can be solved without any programming. That then gives researchers scope for whatever variants of box plots are likely to prove interesting and useful. Cleveland's 1985 variant seems especially worthy of further consideration.

5   Errata to previous column

Marcel Zwahlen helpfully pointed out an inconsistency between an equation on p. 308 and the corresponding Mata code on p. 309 within the previous column (Cox 2009). The equation was incorrect.

$$K = \frac{N^2 - \sum_{i=1}^{I} n_i^2}{\sum_{i=1}^{I} n_i - N q_i}$$

should have been

$$K = \frac{N^2 - \sum_{i=1}^{I} n_i^2}{\sum_{i=1}^{I} (n_i - N q_i)^2}$$

6 References

Baum, C. F. 2009. *An Introduction to Stata Programming*. College Station, TX: Stata Press.

Chen, C., W. Härdle, and A. Unwin, eds. 2008. *Handbook of Data Visualization*. Berlin: Springer.

Cleveland, W. S. 1985. *The Elements of Graphing Data*. Monterey, CA: Wadsworth.

Cox, N. J. 2002. Speaking Stata: On getting functions to do the work. *Stata Journal* 2: 411–427.

———. 2004. Stata tip 6: Inserting awkward characters in the plot. *Stata Journal* 4: 95–96.

———. 2005. Speaking Stata: The protean quantile plot. *Stata Journal* 5: 442–460.

———. 2006. Stata tip 39: In a list or out? In a range or out? *Stata Journal* 6: 593–595.

———. 2007. Stata tip 47: Quantile–quantile plots without programming. *Stata Journal* 7: 275–279.

———. 2009. Speaking Stata: I. J. Good and quasi-Bayes smoothing of categorical frequencies. *Stata Journal* 9: 306–314.

Cox, N. J., and K. Jones. 1981. Exploratory data analysis. In *Quantitative Geography: A British View*, ed. N. Wrigley and R. J. Bennett, 135–143. London: Routledge and Kegan Paul.

Crowe, P. R. 1933. The analysis of rainfall probability: A graphical method and its application to European data. *Scottish Geographical Magazine* 49: 73–91.

Cui, J. 2007. Stata tip 42: The overlay problem: Offset for clarity. *Stata Journal* 7: 141–142.

De Veaux, R. D., P. F. Velleman, and D. E. Bock. 2008. *Stats: Data and Models*. 2nd ed. Boston: Addison–Wesley.

Dümbgen, L., and H. Riedwyl. 2007. On fences and asymmetry in box-and-whiskers plots. *American Statistician* 61: 356–359.

Freedman, D., R. Pisani, and R. Purves. 2007. *Statistics*. 4th ed. New York: W. W. Norton.

Frigge, M., D. C. Hoaglin, and B. Iglewicz. 1989. Some implementations of the boxplot. *American Statistician* 43: 50–54.

Harris, R. L. 1999. *Information Graphics: A Comprehensive Illustrated Reference*. New York: Oxford University Press.

Hoaglin, D. C., F. Mosteller, and J. W. Tukey, eds. 1983. *Understanding Robust and Exploratory Data Analysis*. New York: Wiley.

Hyndman, R. J., and Y. Fan. 1996. Sample quantiles in statistical packages. *American Statistician* 50: 361–365.

Kleiner, B., and T. E. Graedel. 1980. Exploratory data analysis in the geophysical sciences. *Reviews of Geophysics and Space Physics* 18: 699–717.

McGill, R., J. W. Tukey, and W. A. Larsen. 1978. Variations of box plots. *American Statistician* 32: 12–16.

Mitchell, M. N. 2008. *A Visual Guide to Stata Graphics*. 2nd ed. College Station, TX: Stata Press.

Monkhouse, F. J., and H. R. Wilkinson. 1971. *Maps and Diagrams*. London: Methuen.

Spear, M. E. 1952. *Charting Statistics*. New York: McGraw–Hill.

———. 1969. *Practical Charting Techniques*. New York: McGraw–Hill.

Tukey, J. W. 1972. Some graphic and semigraphic displays. In *Statistical Papers in Honor of George W. Snedecor*, ed. T. A. Bancroft and S. A. Brown, 293–316. Ames, IA: Iowa State University Press.

———. 1977. *Exploratory Data Analysis*. Reading, MA: Addison–Wesley.

Velleman, P. F., and D. C. Hoaglin. 1981. *Applications, Basics, and Computing of Exploratory Data Analysis*. Boston: Duxbury.

Wainer, H. 1990. Graphical visions from William Playfair to John Tukey. *Statistical Science* 5: 340–346.

About the author

Nicholas Cox is a statistically minded geographer at Durham University. He contributes talks, postings, FAQs, and programs to the Stata user community. He has also coauthored 15 commands in official Stata. He wrote several inserts in the *Stata Technical Bulletin* and is an editor of the *Stata Journal*.

The Stata Journal (2013)
13, Number 2, pp. 398–400

Speaking Stata: Creating and varying box plots: Correction

Nicholas J. Cox
Department of Geography
Durham University
Durham, UK
n.j.cox@durham.ac.uk

A previous article (Cox 2009) discussed the creation of box plots from first principles, particularly when a box plot is desired that `graph box` or `graph hbox` cannot provide.

This update reports and corrects an error in my code given in that article. The problems are centered on page 484. The question is how to calculate the positions of the ends of the so-called whiskers.

To make this more concrete, the article's example starts with

```
. sysuse lifeexp
. egen upq = pctile(lexp), by(region) p(75)
. egen loq = pctile(lexp), by(region) p(25)
. generate iqr = upq - loq
```

and that holds good.

Given interquartile range (IQR), the position of the end of the upper whisker is that of the largest value not greater than the upper quartile + 1.5 IQR. Similarly, the position of the end of the lower whisker is that of the smallest value not less than the lower quartile − 1.5 IQR.

The problem lines are on page 484:

```
. egen upper = max(min(lexp, upq + 1.5 * iqr)), by(region)
. egen lower = min(max(lexp, loq - 1.5 * iqr)), by(region)
```

This code works correctly if there are no values beyond where the whiskers should end. Otherwise, it yields upper quartile + 1.5 IQR as the position of the upper whisker, but this position will be correct only if there are values equal to that. Commonly, that position will be too high. A similar problem applies to the lower whisker, which commonly will be too low.

More careful code might be

```
. egen upper2 = max(lexp / (lexp < upq + 1.5 * iqr)), by(region)
. egen lower2 = min(lexp / (lexp > loq - 1.5 * iqr)), by(region)
```

That division / may look odd if you have not seen it before in similar examples. But it is very like a common kind of conditional notation often seen,

$$\max(argument \,|\, condition)$$

or

$$\min(argument \,|\, condition)$$

where we seek the maximum or minimum of some argument, restricting attention to cases in which a specified condition is satisfied, or true.

The connection is given in this way. Divide an argument by a logical expression that evaluates to 1 when the expression is true and 0 otherwise. The result is the argument remains unchanged on division by 1 but evaluates as missing on division by 0. In any context where Stata ignores missings, that is what is wanted. True cases are included in the computation, and false cases are excluded.

This "divide by zero" trick appears not to be widely known. There was some publicity within a later article (Cox 2011).

Turning back to the box plots, we will see what the difference is in our example.

```
. tabdisp region, c(upper upper2 lower lower2)
```

Region	upper	upper2	lower	lower2
Eur & C.Asia	79	79	65	65
N.A.	79	79	58.5	64
S.A.	75	75	63	67

Here `upper2` and `lower2` are from the more careful code just given, and `upper` and `lower` are from the code in the 2009 column. The results can be the same but need not be.

Checking Stata's own box plot

```
. graph box lexp, over(region) yli(75 79 64 65 67)
```

shows consistency with the corrected code.

Thanks to Sheena G. Sullivan, UCLA, who identified the problem on Statalist (http://www.stata.com/statalist/archive/2013-03/msg00906.html).

1 References

Cox, N. J. 2009. Speaking Stata: Creating and varying box plots. *Stata Journal* 9: 478–496.

———. 2011. Speaking Stata: Compared with ... *Stata Journal* 11: 305–314.

About the author

Nicholas Cox is a statistically minded geographer at Durham University. He contributes talks, postings, FAQs, and programs to the Stata user community. He has also coauthored 15 commands in official Stata. He wrote several inserts in the *Stata Technical Bulletin* and is an editor of the *Stata Journal*.

The Stata Journal (2009)
9, Number 4, pp. 621–639

Speaking Stata: Paired, parallel, or profile plots for changes, correlations, and other comparisons

Nicholas J. Cox
Department of Geography
Durham University
Durham, UK
n.j.cox@durham.ac.uk

Abstract. Paired, parallel, or profile plots showing the values of two variables may be constructed readily using a combination of `graph twoway` commands. This column explores the principles and practice of such plot-making, considering both wide and long (panel or longitudinal) data structures in which such data may appear. Applications include analysis of change over time or space and indeed any kind of correlation or comparison between variables. Such plots may be extended to show numeric values and associated name information.

Keywords: gr0041, profile plot, parallel coordinates plot, parallel line plot, pairlink diagram, bumps charts, barometer charts, graphics, panel data, longitudinal data, arrows, twoway

1 Introduction

1.1 Graphical comparison of two variables

Comparison of two variables is a graphical problem that arises in many different situations.

The variables are often raw data, but need not be. They could be, for example, sets of summary statistics or quantities calculated from fitting one or more models (residuals, predicted values, figures of merit, etc.). Here the focus is on examples in which the variables, however defined, are recorded on identical (or at least comparable) scales. That restriction is not strong, as it could easily be satisfied by some kind of standardizing or ranking.

The situations could also vary. They include studies of change over time or space, correlations between variables, and other kinds of comparison. However, the graphical problem is much the same.

This column is a sequel to an earlier discussion of graphing agreement and disagreement (Cox 2004). After that column was published, Stata 9 added a set of paired-coordinate commands to `graph twoway`, which makes several pertinent graphs much easier. Cox (2005b) publicized the use of `twoway pcarrow` for graphing changes over time, but the wider possibilities still deserve attention. Although user-written commands are available, the emphasis here is on working out how to plot data yourself with `twoway` from first principles.

1.2 Pairs, parallel coordinates, and profiles

The main kind of graphs under consideration goes under several different names, often emphasizing variously the paired, parallel, or profile character of the plot. You may know yet other names used in your field. Glance ahead to figures 3, 4, 7, and 8 to get a picture of what is being talked about.

Whatever the name, such graphs have a long history. Friendly (2007) drew attention to the use by Guerry (1833) of what many now call a parallel coordinates plot for comparing relative frequency of crimes at different ages. Friendly (2008) adds further nineteenth century examples. Such plots have been the subject of many recent accounts, ranging from introductory (Robbins 2005; Few 2009) through intermediate (Wilkinson 2005; Unwin, Theus, and Hofmann 2006; Cook and Swayne 2007; Chen, Härdle, and Unwin 2008; Theus and Urbanek 2009) to more advanced (Inselberg 2009).

Interaction plots long common in looking at analyses of variance (e.g., Cox [1958]) could be considered as a variation on the main idea.

"Bumps charts" are a version of parallel coordinates plots that often appear in newspapers or on the web. Such charts originally showed changes in rank in series of rowing races held at Oxford and Cambridge and also elsewhere. Given relatively narrow rivers, boats start in single file and crews aim to overtake or "bump" boats in front and not be overtaken from behind. Tufte (1990, 111; 2006, 56) helped publicize bumps charts to a wider readership. The term is now often used beyond its sporting origins.

Another very common variant is widely known as a profile plot. For example, in behavioral research, human or animal subjects may be monitored through time or according to various tests or measures. Data for each subject are plotted as a connected line or profile. That usage is broadly consistent with others. In some Earth or environmental sciences, profiles record variation in properties such as surface altitude along paths or transects in space (e.g., Cox [1990]). Basford and Tukey (1997, 1999) made extensive use of profile plots in a major graphically-based analysis of a plant breeding trial.

Campbell and Kenny (1999) showed how such graphs, which they called pair-link diagrams, could be used in discussing regression artifacts such as regression toward the mean. Wallgren et al. (1996) used the term barometer charts, while Harris (1999) wrote of comparative graphs.

Wilkinson (2005, 314) suggested a distinction: profile plots have a common measurement scale but parallel coordinates plots do not. This distinction is puzzling (why plot at all if the scale is not common in some sense?) and, more crucially, it appears to be neither preached nor practiced widely. To muddy the waters further, further senses of profile plots can be found in the literature. Ramsey and Schafer (2002) use the term for graphs in which each individual is plotted in a separate panel so that profiles are not superimposed. On the other hand, du Toit, Steyn, and Stumpf (1986) refer to both separate and superimposed traces as profiles.

Parallel coordinates and profile plots lend themselves easily to plotting several variables simultaneously, which is a major attraction. Indeed, that is exactly where most treatments start. However, problems with just two variables are sufficiently common to merit detailed attention. If data for only two variables are being plotted, there is usually scope to elaborate the graph by adding information on (say) identifying names or the values themselves. Conversely, a common criticism of such plots (e.g., Venables and Ripley [2002]; Cox [2004]; Young, Valero-Mora, and Friendly [2006]) is that they may become busy and confusing. Naturally, the aim is to avoid, or at least to reduce, such confusion and to turn to other graph forms if they work better for some problems.

2   Treatments for anorexia

2.1   The example data

An interesting dataset that provides a suitable example comes from Hand et al. (1994, 229). They reported some data from Brian S. Everitt on weights of young girls receiving different treatments for anorexia. The weights were said to be in kg, but are clearly in pounds (lb), as McNeil (1996, 57) also commented. The weights are reported for before and after various treatments: cognitive behavioral therapy, control, and family therapy. Hand et al. also comment: "Whichever statistical technique is employed, it is instructive to look at the three scatterplots of after/before." The data are provided with the media for this issue. They are also available at http://www.stat.ucla.edu/data/hand-daly-lunn-mcconway-ostrowski/ANOREXIA.DAT.

2.2   Scatterplot

Let us start with the advice given. Figure 1 is a scatterplot.

```
. use anorexia
. scatter after before before, ms(Oh i) c(. l) lc(none gs12)
> sort(before) yla(, ang(h)) ytitle(after)
> by(treatment, row(1) note("weight, lb") legend(off))
```

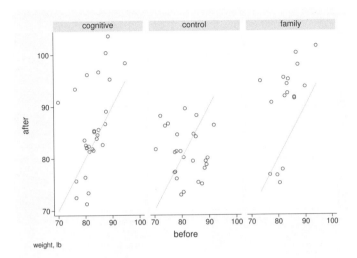

Figure 1: Weights of anorexic girls before and after various treatments. Diagonal lines mark no change in weight.

This scatterplot plots `after` against `before` and adds reference lines $y = x$ by also plotting `before` versus itself. The data are plotted separately by treatment in one row, `by(treatment, row(1))`. The reference lines are plotted with subdued color, `lc(gs12)`, while `sort(before)` ensures that they are plotted smoothly. Other options above tune cosmetic details. Alphabetical order leaves control subjects in the middle, which can be regarded as fortunate.

To get a feel for magnitudes, readers in most countries may like to know that 30 (40, 50) kg are about 66 (88, 110) pounds, spanning the range shown.

The scatterplot does show clearly the broad features of the data. The impressions are that control subjects are about equally divided between weight gainers and losers. Most subjects gained weight with family therapy, but a distinct group lost weight substantially. Weights generally improved with cognitive behavioral therapy, but there were also several exceptions.

2.3   Parallel coordinates plots

A limitation of the scatterplot is that change in weight, after − before, is encoded only indirectly, despite being the response measure of most interest. A parallel coordinates plot is a move toward more direct encoding. For such a plot, we already have the variables `before` and `after` to serve as parallel y coordinates; we just need to construct the corresponding x coordinates. Convention puts `before` to the left of `after` and convenience leads to a choice such as

```
. generate byte one = 1
. generate byte two = 2
```

Sticklers for style and efficiency will appreciate that those variables are produced as `byte`. The names are arbitrary, and we will make sure that graph readers never see them.

As a first stab, we will show paired values by spikes using `twoway pcspike`. Spikes here are straight line segments with no symbol at either end. For more about the alternatives, start with [G-2] **graph twoway** and look at the list of other paired-coordinate graph types. We will shortly look at `twoway pcarrow`. Figure 2 is the result of plotting changes as spikes.

```
. twoway pcspike before one after two,
> xla(1 "before" 2 "after") xtitle("") yla(, ang(h)) ytitle("weight, lb")
```

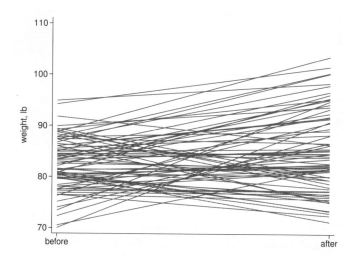

Figure 2: Rough parallel coordinates or profile plot for weights of anorexic girls before and after various treatments

The result is rather busy. Clearly, we want to move on to a graph that separates the different treatments. Figure 3 is the result.

```
. twoway pcspike before one after two,
> xla(1 "   before" 2 "after   ")) xtitle("")
> yla(, ang(h) nogrid) ytitle("weight, lb")
> by(treatment, row(1) noixtick legend(off))
```

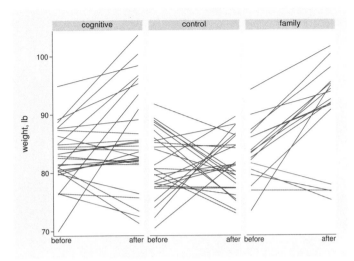

Figure 3: Improved parallel coordinates or profile plot for weights of anorexic girls before and after various treatments

When we do that, we add some small changes. The labels `before` and `after` are pushed inward with extra spaces. The associated ticks do no good, so they are suppressed by the `noixtick` suboption. The grid of horizontal lines is also a distraction given the graph style of line segments. The legend is suppressed, as in figure 1.

We can go further. A basic feature that bears a little emphasis is the contrast between gainers and losers of weight. Two simple choices are to emphasize a minority (the converse can be too loud) and to emphasize any group that is scientifically interesting or practically important. The two choices give the same answer here: stress those who lost weight.

In this column, the distinction is made by line width and grayscale color (Cox 2009). There is also freedom to vary line pattern. You may well have greater freedom yet, say, to choose bolder colors for a presentation.

```
. twoway pcspike before one after two if before <= after, lcolor(gs12) ||
> pcspike before one after two if before > after,
> lw(*1.2) lcolor(gs2) xla(1 "    before" 2 "after    ")
> xtitle("") yla(, nogrid ang(h)) ytitle("weight, lb")
> by(treatment, row(1) note("") legend(off) noixtick)
```

Figure 4 is the result.

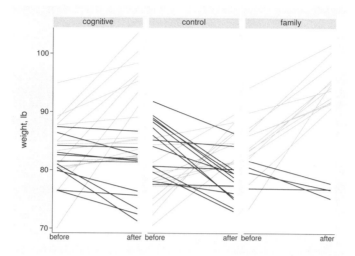

Figure 4: Further improved parallel coordinates or profile plot for weights of anorexic girls before and after various treatments. Girls who lost weight are emphasized.

An obvious but nevertheless common small error is to forget about equal values. That is, > and < usually need to include = on one side or the other. The exceptions are when you really do intend to look only at cases that changed one way or the other and omit unchanged cases from the graph.

Whatever you want to emphasize should be plotted second, that is, on top, so that any overwriting on the graph is in its favor. Thus the trick is to plot spikes for the losers on top of spikes for the gainers, and thicken the lines and darken the color for the losers.

2.4 Parallel line plots

Another plot form that shows after − before even more directly is the parallel line plot (McNeil 1992, 1996; Cox 2004, 2006). The version here capitalizes on the rough equality of treatment numbers, but if groups were very unequal, that could be accommodated otherwise. There are no identifiers or other variables in the published data, so sorting might as well be by weight, either before or after treatment. Any ties should be sorted tidily. A horizontal coordinate can then be the order by weight, except that subtracting the approximate mean rank will center each display.

Weight before is perhaps the more obvious choice.

```
. bysort treatment (before after) : generate order1 = _n - _N/2
```

Readers needing more information on the by: prefix—including the principle that under by:, the built-in variables _n and _N are interpreted within groups—can find a tutorial in Cox (2002).

```
. twoway pcarrow before order1 after order1, pstyle(p1)
> || scatter before order1, pstyle(p1) ms(o)
> xla(none) xtitle("") yla(, ang(h)) ytitle("weight, lb")
> by(treatment, row(1) note("") legend(off))
```

Figure 5 is the result.

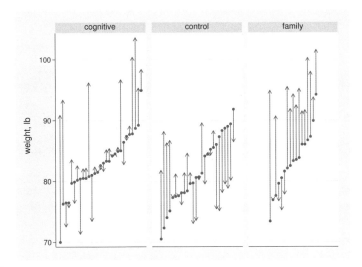

Figure 5: Parallel line plot for weights of anorexic girls before and after various treatments. Point symbols indicate weights before treatment. Arrows point in the direction of change.

We turn to arrows to show change (Cox 2005b) but add marker symbols for the weights before treatment to give slight emphasis to the distribution as a set. Thus each panel is a quantile plot (Cox 2005a), together with added vectors of change. `pstyle(p1)` is a trick to ensure consistent style for both the arrows and the scatterplot. The x axis does have a meaning as rank order, but that remains tacit.

It is also possible to adopt the opposite point of view. We may focus on the end or later result and ask: How did the individuals get to here? In education, for example, you might focus on exit grade-point average or other final achievement measure, as compared with entry data. The code is simply a variation on that just seen. Figure 6 is the result.

```
. bysort treatment (after before) : generate order2 = _n - _N/2
. twoway pcspike before order2 after order2, pstyle(p1)
> || scatter after order2, pstyle(p1) ms(o)
> xla(none) xtitle("") ytitle("weight, lb") yla(, ang(h))
> by(treatment, row(1) note("") legend(off))
```

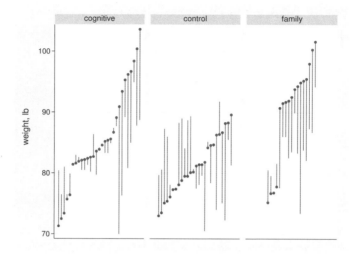

Figure 6: Parallel line plot for weights of anorexic girls before and after various treatments. Point symbols indicate weights after treatment. Spikes indicate the change from before treatment.

You can see one detailed change: spikes were used rather than arrows to avoid the graph becoming too busy around the data points. Clearly, you could have arrows if you wanted them, and you could tweak the arrowhead display to make it just subtly noticeable.

Tastes and judgments will differ, but these two last graphs are my favorites for these data. Other graphs could be shown too, say, box plots (as in McNeil [1996]) or quantile–quantile plots. However, neither of those graph forms respects the pairing in the data. It is true that figures 5 and 6 may not extend easily to series of three or more measurements. Figure 4 would be much more flexible in that regard. However, exploiting the structure of the data to a good end is entirely fair play.

3   Panel or longitudinal structure

Many readers will have been surprised that these data have not yet been described as panel or longitudinal data and treated as such. They will regard using what they recognize as a wide data structure (each girl as one observation) as perverse and prefer a long structure (each girl as two observations). Nothing in the information indicates that the times of measurements before and after were the same, or even equally spaced, so any times assigned are wholly relative and arbitrary. But that corresponds exactly to how the data have been treated so far.

3.1 Mapping to long data structure

If we back-tracked to have only the original data in memory, then there are at least two ways to map the data to long structure.

```
. keep treatment before after
```

Either way, we need an identifier to help Stata keep track, even if we have to invent it ourselves.

```
. generate id = _n
```

The first way to restructure, and probably the better known among Stata users, is to `reshape`. Stata will not regard `before` and `after` as cognate unless their names share a prefix:

```
. rename before weight1
. rename after weight2
. reshape long weight, i(id) j(time)
```

The second way would be to use `stack`:

```
. stack id before treatment id after treatment, into(id weight treatment) clear
. rename _stack time
```

Regardless of how you restructure, and of whether you declare the panel data as such using `tsset` or `xtset`, it is a good idea to sort the data now:

```
. sort id time
```

3.2 Graphs from long data structure

Turning now to graphs, first note that `xtline` is a dead end for our problem, because it does not support a `by()` option. We can easily work from first principles instead.

```
. line weight time, c(L) xla(1 "  before" 2 "after    ") xtitle("")
> yla(, ang(h)) ytitle("weight, lb")
> by(treatment, row(1) note("") noixtick)
```

produces the spitting image of figure 3. `c(L)` here is an old Stata trick: exactly the same syntax has carried over from the old graphics before Stata 8. `c(L)` says "join the data if and only if the x axis variable is increasing". The previous `sort id time` means that `time` goes 1 2 1 2 1 2 and so forth for identifiers 1 2 3 and so forth, and so joining of points takes place only within panels, as 1 increases to 2, and not between panels, as 2 decreases back to 1. Thus that one option ensures that data points are connected only for each individual girl.

How do we separate weight losers and gainers with this data structure? Consider the command

```
. by id: generate byte falling = weight[2] < weight[1]
```

What happens with this command depends delicately on the previous `sort id time`, so for clarity let us combine the two commands:

```
. bysort id (time): generate byte falling = weight[2] < weight[1]
```

The trick here is very Stataish. Some readers may smile, recognizing an old friend. The tutorial earlier mentioned (Cox 2002) rehearses the basics. The key point is that under `by:`, subscripts such as `[1]` and `[2]` are interpreted within the groups defined by `by:`. Here that means groups defined by distinct values of `id`, the panels or individual girls. The panels for this dataset are balanced with precisely two observations each, so everything is about as simple as it could be.

With this sort order, the first observation in each panel, subscripted `[1]`, is the first in `time`, and the second, `[2]`, is the second in `time`. Therefore, for both observations in each panel, the new variable created with the value for that panel of

```
weight[2] < weight[1]
```

which is true, evaluated as 1, if `weight` fell from time 1 to time 2 and false, evaluated as 0, otherwise.

Let us stay with this syntax briefly and note some further implications. Suppose first that the data were organized in panels of three or more observations. Then the expression just above would still be evaluated using the values for the first two observations in each panel and that would be done also for observations other than the first two. There is no rule that expressions evaluated must refer to any data in the current observation. If the reference was just to `weight`, then Stata would always use the value of `weight` in each observation, but explicit subscripts override that kind of reference.

Suppose now that there was only one observation in a panel. Then the reference to `weight[2]` remains legal but Stata would not find a corresponding value and would return missing, in this problem, numeric missing (`.`). The expression `weight[2] < weight[1]` would then return false unless the value of `weight[1]` was itself missing. In this example, no great harm would ensue, but in other problems you might be bitten, so watch out.

```
. bysort id (time): generate byte falling = weight[2] < weight[1] if _N == 2
```

would trap this particular problem.

Back to the graphics: With our new indicator variable, `falling`, we have the means to separate weight losers and gainers.

One technique is to superimpose graphs for the subset `if falling` and the subset `if !falling`, as used before for figure 4. Another technique is to use `separate` first, which we will show as a variation. Cox (2005c) gave another example of the use of `separate` for scatterplot. The result is a replica of figure 4.

```
. separate weight, by(falling)
. line weight? time, c(L L) lp(solid ..) lc(gs12 gs2)
> xla(1 "  before" 2 "after   ") xtitle("") yla(, ang(h)) ytitle("weight, lb")
> by(treatment, row(1) note("") legend(off) noixtick)
```

`weight?` here is a wildcard that catches the variables `weight0` and `weight1` produced by `separate`. To replicate figure 4, we need to spell out that the line patterns for the two variables are the same; alternatively, we could have different line patterns if we so wished.

However, replicating figures 5 and 6 would be easier with a restructuring back to the data structure we started with, so in one sense, we now close a loop with this example.

4   Big rivers

4.1   The example data

We turn now to a very different example in which two measured variables are on quite different scales, so that we choose to compare them using ranks, and in which we also have names that we wish to see.

Allen (1997, 136–139) gave data on 97 of the world's largest rivers. We focus on basin (catchment or watershed) area and mean discharge, the mean volume of water per unit time leaving the river basin. As an aside, note that complete databases for even the largest rivers remain elusive and that definitions and measurements of these and other quantities are highly problematic. See, for example, the often different data listings of Gleick (1993) or Shiklomanov and Rodda (2003).

The data are provided with the media for this issue.

```
. use rivers, clear
```

4.2   Paired plots with names

The initial stimulus for writing this particular column was seeing some attractive displays produced by Fry (2008), broadly similar to what we are going to see. In essence, they are modern equivalents of Guerry's displays (Guerry 1833).

For our illustration, we select the 25 largest rivers according to basin area.

```
. gsort -area
. keep in 1/25
(72 observations deleted)
```

```
. list name area discharge
```

		name	area	discha~e
1.	Amazon		6150	200000
2.	Zaire (Congo)		3700	40900
3.	Mississippi		3344	18400
4.	Nile		2715	317
5.	Parana		2600	18000
6.	Yenisei		2580	17800
7.	Ob		2500	12200
8.	Lena		2430	16200
9.	Yangtze		1940	28500
10.	Amur		1855	10300
11.	Mackenzie		1448	9830
12.	Zambezi		1400	6980
13.	Volga		1350	8400
14.	St Lawrence		1185	14300
15.	Niger		1112.7	6020
16.	Shatt al Arab		1050	1460
17.	Ganges		980	11600
18.	Yellow (Huang He)		980	1550
19.	Indus		960	7610
20.	Orinoco		945	34900
21.	Murray		910	698
22.	Chari		880	1320
23.	Yukon		855	6180
24.	Danube		815	6660
25.	Mekong		810	14900

The area of the Niger basin for some reason is recorded with an extra decimal place. We will round for display purposes.

```
. replace area = round(area)
```

For the vertical coordinates, we need ranks on area and discharge. Given our previous sort, the first is immediately accessible as the observation number. For the second, we use egen, rank(). Note the use of a negative sign to ensure consistent ranking so that the largest is first.

```
. generate rank1 = _n
. egen rank2 = rank(-discharge)
```

It so happens that there are no ties. egen, rank() has a unique option for that situation.

The horizontal coordinates we can set with small integers, as before:

```
. generate byte one = 1
. generate byte two = 2
```

I am going to give the rest of the code all at once and then comment once you have seen the display. This code is, not surprisingly, a cleaned-up version after various small experiments. My experience is simply that you should get a rough version up and running and then improve it step by step.

```
. generate left = 0.4

. twoway pcspike rank1 one rank2 two,
> xla(none) xsc(noline r(0.3 2.3)) xtitle("")
> ysc(r(-1 .) reverse off) yla(, nogrid)
> || scatter rank1 one, mla(area) mlabpos(9) ms(none)
> || scatter rank2 two, mla(discharge) mlabpos(3) ms(none)
> || scatter rank1 left, mla(name) mlabpos(3) ms(none)
> text(-0.5 1 "area, 000 sq.km") text(-0.5 2 "discharge, cu.m/s")
> legend(off) graphregion(color(white))
```

Figure 7 shows the result.

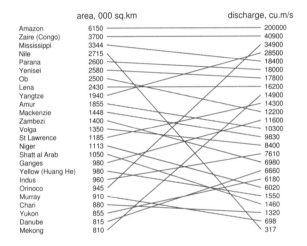

Figure 7: Paired display ranking 25 big rivers by both area and mean discharge

Comments now:

1. The spikes are drawn by `twoway pcspike`, as is now familiar.

2. The values of `area`, as just rounded slightly, are displayed to the left using a `scatter` with vertical coordinate the rank, horizontal coordinate 1, invisible marker symbol, and the values as marker labels. `mlabpos(9)` puts the labels at the 9 PM position, i.e., to the left of where the marker symbols would have been visible.

3. Similarly, the values of `discharge` are displayed to the right using a `scatter` with vertical coordinate the rank, horizontal coordinate 2, invisible marker symbol, and

the values as marker labels. `mlabpos(3)` puts the labels at the 3 PM position, i.e., to the right of where the marker symbols would have been visible.

4. The river names are shown using the same device as in 2 and 3. The horizontal position of 0.4 is the result of experiment. Clearly, longer or shorter names, or different judgments about spacing, would mean different values.

5. Column headers are added through `text()` options.

6. Given the extra material on the margins of the graph, we need to stretch axis limits using `xscale()` and `yscale()`.

7. We want rank 1 at the top and rank 25 at the bottom, so spell out `ysc(reverse)`.

8. The rest of the code consists of subtracting stuff that is unnecessary or would be a distraction:

```
xla(none) xsc(noline) xtitle("") ysc(off) yla(, nogrid) legend(off) ///
    graphregion(color(white))
```

Bear in mind that what is shown just above is specific to the `sj` scheme. If other graph schemes had been used, the code might have been slightly different.

4.3 Other possibilities

The main point of the example is to show something of what is possible. Other possibilities now open up in turn. For example, we could highlight particular observations or groups of observations. Someone might want to emphasize that the Nile is anomalous in having high area but low discharge, a result of arid climate in its lower parts and much extraction of water for human uses, including irrigation. Figure 8 shows how this might appear. The particular change here of thickening one spike was made in the Graph Editor without needing to work out the command-based logic.

Some might want to tweak the presentation by using different justifications (left, right, centered) of the numeric values. Others might want to subdivide rising and falling groups, those with higher ranks on one variable than another. Fry (2008) gives some good examples of this style.

Now that ranks have been mentioned, we should spell out what will be intuitive: the connection between these graphs and rank correlation. At one extreme, perfect coincidence of ranks corresponds to both perfect positive rank correlation and a graph with no crossings of lines. For details on the relationship between line crossings and rank correlation, start with Fisher (1983). Except in extreme cases, anyone wanting to see a rank correlation will still find it faster to use the corresponding Stata command.

People who work with panel or longitudinal data have wrestled with the problem that graphs with many panels superimposed can become too busy. For many researchers, the datasets looked at here are just toy examples: their sample sizes are hundreds, thousands, or millions. The common affectionate reference to spaghetti plots understates the problem, because graphs can be much more entangled even than spaghetti ever gets.

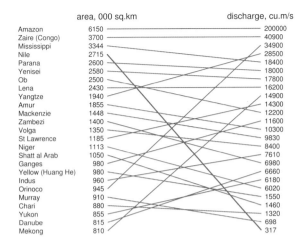

Figure 8: Paired display ranking 25 big rivers by both area and mean discharge. Note how the Nile has very low discharge considering its area.

One standard idea is to plot the mass of curves in subdued form or as unconnected data points and plot some kind of regression or smooth summary on top (e.g., Singer and Willett [2003]). This can solve the problem of not being able to see the wood for the trees, as the old English cliché has it. However, how individual trees change is also of central concern.

Diggle et al. (2002, 38–39) suggest an interesting compromise. Choose the individuals that on some characteristic (say, their median residual from the overall mean or smooth curve) lie at certain selected quantiles throughout the distribution. For concreteness, these might be the minimum, the 5th, 10th, 25th, 50th, 75th, 90th, and 95th percentiles, and the maximum. Thus if these individual curves were plotted, something of the range and style of variability would remain evident in the plot and not be submerged in a gray mess of data points. Although this idea is not explored here, its application in Stata would rest on the devices exemplified in this column.

5   Conclusion

In graphics as in the rest of statistical science, the very simplest ideas are often the best. Plotting data for two variables in parallel is a beautifully simple idea made easy to implement by Stata's paired-coordinate graphs. Only experiment indicates what works best with particular datasets, but it is important to remember that other possibilities lie beyond the scatter or time-series plot.

6 Acknowledgment

Rino Bellocco drew my attention to Campbell and Kenny (1999).

7 References

Allen, P. A. 1997. *Earth Surface Processes*. Oxford: Blackwell Science.

Basford, K. E., and J. W. Tukey. 1997. Graphical profiles as an aid to understanding plant breeding experiments. *Journal of Statistical Planning and Inference* 57: 93–107.

————. 1999. *Graphical Analysis of Multiresponse Data: Illustrated with a Plant Breeding Trial*. Boca Raton, FL: Chapman & Hall/CRC.

Campbell, D. T., and D. A. Kenny. 1999. *A Primer on Regression Artifacts*. New York: Guilford Press.

Chen, C., W. Härdle, and A. Unwin, eds. 2008. *Handbook of Data Visualization*. Berlin: Springer.

Cook, D., and D. F. Swayne. 2007. *Interactive and Dynamic Graphics for Data Analysis: With R and GGobi*. New York: Springer.

Cox, D. R. 1958. *Planning of Experiments*. New York: Wiley.

Cox, N. J. 1990. Hillslope profiles. In *Geomorphological Techniques*, ed. A. S. Goudie, 92–96. London: Unwin Hyman.

————. 2002. Speaking Stata: How to move step by: step. *Stata Journal* 2: 86–102.

————. 2004. Speaking Stata: Graphing agreement and disagreement. *Stata Journal* 4: 329–349.

————. 2005a. Speaking Stata: The protean quantile plot. *Stata Journal* 5: 442–460.

————. 2005b. Stata tip 21: The arrows of outrageous fortune. *Stata Journal* 5: 282–284.

————. 2005c. Stata tip 27: Classifying data points on scatter plots. *Stata Journal* 5: 604–606.

————. 2006. Assessing agreement of measurements and predictions in geomorphology. *Geomorphology* 76: 332–346.

————. 2009. Stata tip 78: Going gray gracefully: Highlighting subsets and downplaying substrates. *Stata Journal* 9: 499–503.

Diggle, P. J., P. Heagerty, K.-Y. Liang, and S. L. Zeger. 2002. *Analysis of Longitudinal Data*. 2nd ed. Oxford: Oxford University Press.

du Toit, S. H. C., A. G. W. Steyn, and R. H. Stumpf. 1986. *Graphical Exploratory Data Analysis*. New York: Springer.

Few, S. 2009. *Now You See It: Simple Visualization Techniques for Quantitative Analysis*. Oakland, CA: Analytics Press.

Fisher, N. I. 1983. Graphical methods in nonparametric statistics: A review and annotated bibliography. *International Statistical Review* 51: 25–38.

Friendly, M. 2007. A.-M. Guerry's *Moral Statistics of France*: Challenges for multivariable spatial analysis. *Statistical Science* 22: 368–399.

———. 2008. The golden age of statistical graphics. *Statistical Science* 23: 502–535.

Fry, B. 2008. *Visualizing Data: Exploring and Explaining Data with the Processing Environment*. Sebastopol, CA: O'Reilly.

Gleick, P. H., ed. 1993. *Water in Crisis: A Guide to the World's Fresh Water Resources*. New York: Oxford University Press.

Guerry, A.-M. 1833. *Essai sur la Statistique Morale de la France*. Paris: Crochard.

Hand, D. J., F. Daly, A. D. Lunn, K. J. McConway, and E. Ostrowski, eds. 1994. *A Handbook of Small Data Sets*. London: Chapman & Hall.

Harris, R. L. 1999. *Information Graphics: A Comprehensive Illustrated Reference*. New York: Oxford University Press.

Inselberg, A. 2009. *Parallel Coordinates: Visual Multidimensional Geometry and Its Applications*. New York: Springer.

McNeil, D. 1992. On graphing paired data. *American Statistician* 46: 307–311.

———. 1996. *Epidemiological Research Methods*. Chichester, UK: Wiley.

Ramsey, F. L., and D. W. Schafer. 2002. *The Statistical Sleuth: A Course in Methods of Data Analysis*. 2nd ed. Pacific Grove, CA: Duxbury.

Robbins, N. M. 2005. *Creating More Effective Graphs*. Hoboken, NJ: Wiley.

Shiklomanov, I. A., and J. C. Rodda, eds. 2003. *World Water Resources at the Beginning of the 21st Century*. Cambridge: Cambridge University Press.

Singer, J. D., and J. B. Willett. 2003. *Applied Longitudinal Data Analysis: Modeling Change and Event Occurrence*. Oxford: Oxford University Press.

Theus, M., and S. Urbanek. 2009. *Interactive Graphics for Data Analysis: Principles and Examples*. Boca Raton, FL: Chapman & Hall/CRC.

Tufte, E. R. 1990. *Envisioning Information*. Cheshire, CT: Graphics Press.

———. 2006. *Beautiful Evidence*. Cheshire, CT: Graphics Press.

Unwin, A., M. Theus, and H. Hofmann. 2006. *Graphics of Large Datasets: Visualizing a Million*. New York: Springer.

Venables, W. N., and B. D. Ripley. 2002. *Modern Applied Statistics with S*. 4th ed. New York: Springer.

Wallgren, A., B. Wallgren, R. Persson, U. Jorner, and J.-A. Haaland. 1996. *Graphing Statistics and Data: Creating Better Charts*. Newbury Park, CA: Sage.

Wilkinson, L. 2005. *The Grammar of Graphics*. 2nd ed. New York: Springer.

Young, F. W., P. M. Valero-Mora, and M. Friendly. 2006. *Visual Statistics: Seeing Data with Dynamic Interactive Graphics*. Hoboken, NJ: Wiley.

About the author

Nicholas Cox is a statistically minded geographer at Durham University. He contributes talks, postings, FAQs, and programs to the Stata user community. He has also coauthored 15 commands in official Stata. He wrote several inserts in the *Stata Technical Bulletin* and is an editor of the *Stata Journal*.

The Stata Journal (2010)
10, Number 1, pp. 143–151

Speaking Stata: The statsby strategy

Nicholas J. Cox
Department of Geography
Durham University
Durham City, UK
n.j.cox@durham.ac.uk

Abstract. The `statsby` command collects statistics from a command yielding r-class or e-class results across groups of observations and yields a new reduced dataset. `statsby` is commonly used to graph such data in comparisons of groups; the `subsets` and `total` options of `statsby` are particularly useful in this regard. In this article, I give examples of using this approach to produce box plots and plots of confidence intervals.

Keywords: gr0045, statsby, graphics, groups, comparisons, box plots, confidence intervals

1 Introduction

Datasets are often subdivided at one or more levels according to some kind of group structure. Statistically minded researchers are typically strongly aware of the need for, and the value of, comparisons between patients, hospitals, firms, countries, regions, sites, or whatever the framework is for collecting and organizing their data. Indeed, for many people, that kind of comparison is at the heart of what they do daily within their research.

Stata supports separate group analyses in various ways. Perhaps the most well-known and important is the `by:` construct, a subject of one of the earliest *Speaking Stata* columns (Cox 2002). This column focuses on [D] **statsby**, a command that until now has received only passing mention in *Speaking Stata* (Cox 2001, 2003). The main idea of `statsby` is simple and it is well documented. However, experience on Statalist and elsewhere indicates that many users who would benefit from `statsby` are unaware of its possibilities. The extra puff of publicity here goes beyond the manual entry in stressing its potential for graphical comparisons.

Focusing exclusively on `statsby` is not intended as a denial that there are other solutions to the same, or related, problems. The work of Newson (1999, 2000, 2003) is especially notable in this regard and goes beyond the singular purpose explored here.

2 The main idea

The main idea of `statsby` is that it offers a framework, not only for automating separate analyses for each of several groups, but also for collating the results. The effect is to relieve users of much of the tedious organizing work that would be needed otherwise. The default mode of operation is that `statsby` overwrites the original dataset, subdivided in some way, with a reduced dataset with just one observation for each group. The `saving()` option, however, permits results to be saved on the side so that the original dataset remains in memory.

A common and essentially typical example of applying `statsby` is that a panel dataset containing one or more observations for each panel would be reduced to a dataset with precisely one observation for each panel. Those observations contain panel identifiers together with results for each panel, usually e- or r-class results from some command. Because there is no stipulation that the command called is an official command, there is scope for users to write their own programs leaving such results in their wake and thus to automate essentially any kind of calculation.

`statsby` does not support graphs directly, but the implications for graphics are immediate. Graphics for groups imply the collation of group results followed by graphing operations. Using `statsby` can reduce the problem to just the second of these two, subject as usual to minor questions of titling, labeling, and so forth.

3   Box plots for all possible subsets

I will not recapitulate the details of the manual entry, which those unfamiliar with the command can read for themselves. Rather, I will underline the value of `statsby` by showing how it makes several graphical tasks much easier.

Variants on box plots remain popular in statistical science. In an earlier column (Cox 2009), I underlined how `graph twoway` allows your own alternatives if ever the offerings of `graph box` or `graph hbox` are not quite suitable.

Let us pick up that theme and give it a new twist. The `subsets` option of `statsby` makes easy a division into all possible subsets of a dataset. That can be useful so long as you remember enough elementary combinatorics to avoid trying to produce an impracticable or impossible graph.

We will use the `sj` scheme standard for the *Stata Journal* and `auto.dta` bundled with Stata.

```
. set scheme sj
. sysuse auto
(1978 Automobile Data)
```

A small piece of foresight—benefiting from the hindsight given by earlier attempts excised from public view—is now to save a variable label that would otherwise disappear on reduction. In this example, we could also just type in the label or some other suitable text afterward. But if you try something similar yourself, particularly if you want to automate the production of several graphs, the small detail of saving text you want as a graph title may avoid some frustration.

```
. local xtitle "`: var label mpg´"
```

Our call to `statsby` spells out that we want five quantiles, the median, two quartiles, and two extremes. `summarize, detail` does that work. Because `summarize` is an r-class command, we need to look up the codes used, either by reverse engineering from the results of `return list` or by looking at the command help or the manual entry.

The principle with an e-class command is identical, except that we would reverse engineer from `ereturn list`. If this detail on r- and e-class results goes beyond your present familiarity, start at `help saved results` and follow the documentation pointers there if and as desired.

We are subdividing `auto.dta` by the categorical variables `foreign` and `rep78`, but with a twist given by the `subsets` option. Another useful option—in practice, probably even more useful—is to use the `total` option to add results for the whole set.

```
. statsby p50=r(p50) p25=r(p25) p75=r(p75) min=r(min) max=r(max),
> by(foreign rep78) subsets total: summarize mpg, detail
(running summarize on estimation sample)
       command:  summarize mpg, detail
           p50:  r(p50)
           p25:  r(p25)
           p75:  r(p75)
           min:  r(min)
           max:  r(max)
            by:  foreign rep78
Statsby subsets
────┼─── 1 ──┼─── 2 ──┼─── 3 ──┼─── 4 ──┼─── 5
..............
```

Let us look at the results. To emphasize the key point: This is a reduced dataset and the original dataset is gone, although overwriting is avoidable through `saving()`. We have results for all combinations of `foreign` and `rep78` that exist in the data; for all categories of `foreign` and for all categories of `rep78`; and for all observations.

```
. list
```

	foreign	rep78	p50	p25	p75	min	max
1.	Domestic	1	21	18	24	18	24
2.	Domestic	2	18	16.5	23	14	24
3.	Domestic	3	19	16	21	12	29
4.	Domestic	4	18	15	21	14	28
5.	Domestic	5	32	30	34	30	34
6.	Domestic	.	19	16	22	12	34
7.	Foreign	3	23	21	26	21	26
8.	Foreign	4	25	23	25	21	30
9.	Foreign	5	25	18	35	17	41
10.	Foreign	.	25	21	28	17	41
11.	.	1	21	18	24	18	24
12.	.	2	18	16.5	23	14	24
13.	.	3	19	17	21	12	29
14.	.	4	22.5	18	25	14	30
15.	.	5	30	18	35	17	41
16.	.	.	20	18	25	12	41

Plotting that data directly would produce a reasonable working graph. Largely as a matter of personal taste, I chose to reorganize and edit the data slightly to get something more attractive. First, I wanted all two-group categories together, then all one-group categories, and then all the data. The number of groups in each category is the complement of the number of missing values of the first two variables in each observation or row, so that can be calculated by counting missing values in each row and `sorting` accordingly. The `stable` option minimizes departure from the present sort order.

```
. egen order = rowmiss(foreign rep78)
. sort order, stable
```

To get group labels, I combine the value labels (where used) and the values (otherwise) with `egen`'s `concat()` function, and I remove the periods indicating missing and any marginal spaces:

```
. egen label = concat(foreign rep78), decode p(" ")
. replace label = trim(subinstr(label, ".", "", .))
(8 real changes made)
```

The total category for results for all observations deserves due prominence:

```
. replace label = "Total" in L
(1 real change made)
```

The final detail of preparation is to use a couple of helper programs to set up one axis variable with gaps and to map the values in the `label` variable to the value labels of that axis variable:

```
. seqvar x = 1/5 7/9 11/12 14/18 20
. labmask x, values(label)
```

For more details on seqvar and labmask, see Cox (2008).

Now we assemble the box plot from ingredients produced by members of the twoway family. rspike draws spikes between each quartile and each tailward extreme. rbar draws boxes between the quartiles. scatter draws point symbols showing the medians. The result is shown in figure 1. At this point, we use the title carefully stored in a local macro before the call to statsby.

```
. twoway rspike min p25 x, horizontal bcolor(gs12) ||
> rspike p75 max x, horizontal bcolor(gs12) ||
> rbar p25 p75 x, horizontal barw(0.8) bcolor(gs12) ||
> scatter x p50, ms(O) yla(1/5 7/9 11/12 14/18 20, val nogrid noticks ang(h))
> legend(off) ysc(reverse) xtitle(`xtitle')
```

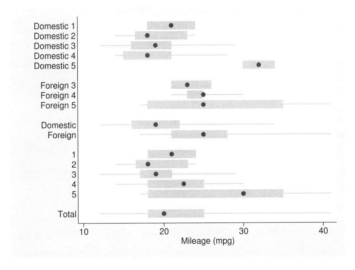

Figure 1: All subsets box plot of mileage for 78 cars by domestic or foreign origin, repair record in 1978, and combinations thereof. Spikes extend to extremes, boxes show quartiles, and circles show medians.

Beyond question, the statistical and stylistic choices here of what to show and how to show it are all arguable and variable. However, that is not the main point. Rather, you should appreciate how statsby with its subsets and total options made a different kind of plot much easier.

4   Confidence-interval plots

Plotting confidence intervals for some group statistic, such as the mean, is another common application. The basic trick, which now starts to look fairly obvious, is to use a command such as `ci` (see [R] **ci**) under the aegis of `statsby` to produce a reduced dataset that is then ready for graphics.

We read in the U.S. National Longitudinal Survey data available from Stata's web site. We will look at the relationship between wage (on an adjusted logarithmic scale) and highest education grade.

```
. webuse nlswork, clear
(National Longitudinal Survey.  Young Women 14-26 years of age in 1968)
```

Saving the variable label is a trick we saw before. At worst, it does no harm.

```
. local ytitle "`: var label ln_wage'"
```

`ci` is our workhorse. The default gives 95% confidence intervals, but evidently other choices may suit specific purposes. As the dataset is panel data, we need to decide how far to respect its structure. One of several possible approaches is to select one observation from each panel randomly (but reproducibly). In addition to estimates and confidence intervals, we save the sample sizes, which are a key part of the information.

```
. set seed 2803
. generate rnd = runiform()
. bysort idcode (rnd): generate byte select = _n == 1
. statsby mean=r(mean) ub=r(ub) lb=r(lb) N=r(N) if select, by(grade) clear:
> ci ln_wage
(running ci on estimation sample)
        command:  ci ln_wage if select
           mean:  r(mean)
             ub:  r(ub)
             lb:  r(lb)
              N:  r(N)
             by:  grade
Statsby groups
——+—— 1 ——+—— 2 ——+—— 3 ——+—— 4 ——+—— 5
(2 missing values generated)
..................
```

Here the grades run over all the integers 0/18, but `levelsof` (see [P] **levelsof**) simplifies capture for later graphical use of all values that occur, especially in more complicated cases.

```
. levelsof grade, local(levels)
0 1 2 3 4 5 6 7 8 9 10 11 12 13 14 15 16 17 18
```

A basic plot is now at hand. Using `scatter` for the means and `rcap` for the intervals themselves is widely conventional. A delicate detail is that means on top of intervals look better than the converse. The result is shown in figure 2.

```
. twoway rcap ub lb grade || scatter mean grade, yti(`ytitle') legend(off)
> subtitle(95% confidence intervals for mean, place(w)) xla(`levels')
```

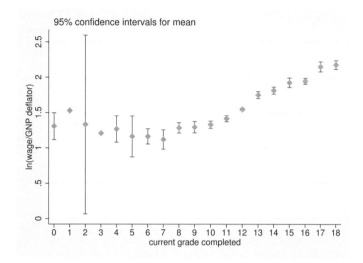

Figure 2: Graph of 95% confidence intervals of mean adjusted log wage by education grade.

The graph can be improved in various minor cosmetic ways and also by showing sample sizes. After some experimenting, the method for the latter was refined to showing sizes as marker labels on a horizontal line. Horizontal alignment of those labels would have been preferable, except that they then would run into one another. Exchanging axes so that grade is plotted vertically seems too awkward for this kind of data. In other circumstances, exchanging axes might well be a good idea. Thus the vertical alignment here is regarded as the lesser of two evils. Figure 3 shows the result.

```
. generate where = 2.7
. twoway rcap ub lb grade || scatter mean grade, yti(`ytitle') legend(off)
> subtitle(95% confidence intervals for mean, place(w)) xla(`levels')
> || scatter where grade, ms(none) mla(N) mlabangle(v) mlabpos(0) ysc(r(. 2.8))
> yla(0(.5)2.5, ang(h))
```

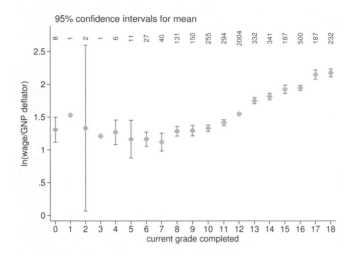

Figure 3: Graph of 95% confidence intervals of mean adjusted log wage by education grade. Text labels show sample sizes at each grade.

5 Conclusions

This column has promoted one simple idea, using `statsby` to prepare a reduced dataset for subsequent graphing. Its `subsets` and `total` options allow useful variations on the default. You might still need to do some further work to get a good graph, but the overall labor is nevertheless likely to be much reduced. The method is widely applicable in so far as any calculation can be represented by a program as yielding r-class or e-class results.

6 Acknowledgments

Vince Wiggins planted the immediate seed for this column with a single cogent remark. Martin Weiss has been an energetic proponent of `statsby` on Statalist.

7 References

Cox, N. J. 2001. Speaking Stata: How to repeat yourself without going mad. *Stata Journal* 1: 86–97.

————. 2002. Speaking Stata: How to move step by: step. *Stata Journal* 2: 86–102.

————. 2003. Speaking Stata: Problems with tables, Part I. *Stata Journal* 3: 309–324.

———. 2008. Speaking Stata: Between tables and graphs. *Stata Journal* 8: 269–289.

———. 2009. Speaking Stata: Creating and varying box plots. *Stata Journal* 9: 478–496.

Newson, R. 1999. dm65: A program for saving a model fit as a dataset. *Stata Technical Bulletin* 49: 2–6. Reprinted in *Stata Technical Bulletin Reprints*, vol. 9, pp. 19–23. College Station, TX: Stata Press.

———. 2000. dm65.1: Update to a program for saving a model fit as a dataset. *Stata Technical Bulletin* 58: 2. Reprinted in *Stata Technical Bulletin Reprints*, vol. 10, p. 7. College Station, TX: Stata Press.

———. 2003. Confidence intervals and p-values for delivery to the end user. *Stata Journal* 3: 245–269.

About the author

Nicholas Cox is a statistically minded geographer at Durham University. He contributes talks, postings, FAQs, and programs to the Stata user community. He has also coauthored 15 commands in official Stata. He wrote several inserts in the *Stata Technical Bulletin* and is an editor of the *Stata Journal*.

The Stata Journal (2010)
10, Number 4, pp. 670–681

Speaking Stata: Graphing subsets

Nicholas J. Cox
Department of Geography
Durham University
Durham, UK
n.j.cox@durham.ac.uk

Abstract. Graphical comparison of results for two or more groups or subsets can be accomplished by way of subdivision, superimposition, or juxtaposition. The choice between superimposition (several groups in one panel) and juxtaposition (several groups in several panels) can require fine discrimination: while juxtaposition increases clarity, it requires mental superimposition to be most effective. Discussion of this dilemma leads to exploration of a compromise design in which each subset is plotted in a separate panel, with the rest of the data as a backdrop. Univariate and bivariate examples are given, and associated Stata coding tips and tricks are commented on in detail.

Keywords: gr0046, graphics, subdivision, superimposition, juxtaposition, quantile plots, Gumbel distribution, scatterplots

1 Introduction

A common graphical problem—indeed for many researchers the key graphical problem— is to compare results for two or more groups or subsets of some larger group or set. Results might be measured responses, absolute or relative frequencies, summary statistics, parameter estimates, model figures of merit, or whatever else is worth plotting. We might seek comparisons according to treatment, disease, gender, ethnicity, industry, product, habitat, land use, area, time period, and so forth: you can multiply examples for yourself.

Various strategies, elementary but also fundamental, recur repeatedly in plotting results for different groups.

Subdivision of some whole is the principle behind pie charts, stacked bar charts, and layered area plots. Each group is represented by its own share, relative or absolute as the case may be.

Superimposition of differing points or lines is the principle behind scatterplots, line plots, and other plots in which different groups are denoted by (for example) distinct marker symbols, marker colors, line patterns, or line colors.

Juxtaposition of separate subpanels or panels within a display conveys group comparisons by what Tufte (2001) called small multiples: the same basic design is repeated for each subset, and possibly also for the total set. Other terms in use are trellis and lattice graphics, multipanel graph, and panel charts (Robbins 2010).

All these ideas are staples within statistical and scientific graphics, and their advantages and disadvantages have been much discussed in many texts and articles. The books of Tufte (1990, 1997, 2001, 2006) and Cleveland (1993, 1994) remain my own favorites as overviews of the field. Despite that large literature, with many preachers and many precepts, it often seems that only experimentation will show which idea is most effective for a particular dataset.

Subdivision is the strategy likely to be first encountered in graphical education through the pies and bars widely met in childhood. However, only some graphical problems reduce to comparing fractions of a whole.

Superimposition promises the advantage of a common scale that can be used for comparison, yet in practice superimposition can mean confusion. The mixture of different elements may appear mostly as a mess in which patterns are difficult to discern. Tangled line plots are often referred to as spaghetti plots, not always with affection or admiration. Some other term (muesli plots?) seems needed for classified scatterplots that convey much detail but in which systematic differences are hard to decipher.

Juxtaposition—in Stata terms often effected by a `by()` or `over()` option—provides separation, which clarifies what is being compared, sometimes at the expense of making that comparison harder work. To work well, juxtaposition still requires mental superimposition. Judging the fine structure of differences between adjacent panels can be difficult enough, and judging differences between panels at opposite ends of a display is evidently even more difficult.

In this column, we will look at a combination of superimposition and juxtaposition, in which subsets are shown separately, but in every case the set as a whole acts as a backdrop. An earlier Stata tip (Cox 2009) emphasized the notion that the whole of the data may serve as a graphical substrate for a particular subset. Repeating information might be criticized as redundant, but rather the idea is that repetition provides reinforcement. Consider a dictum of Tufte (1990, 37): "Simplicity of reading derives from the context of detailed and complex information, properly arranged. A most unconventional design strategy is revealed: to clarify, add detail."

In your own work, you are likely to be able to use color in your talks and possibly also within your reports or even published papers. However, many journals still prohibit or inhibit anything other than black and white and what shades of gray lie between. In this column, we follow such a restriction, using only contrasts discernible by varying gray scale.

2   A univariate example

2.1   Annual maximum windspeeds

A first example concerns data on annual maximum windspeeds for various places in the southeastern United States. My source is Hosking and Wallis (1997, 31); their source is Simiu, Changery, and Filliben (1979). A recipe that has long been a standard in the statistics of extremes is to focus on the maximums of a variable in each of several blocks of time. A year is a natural block for meteorology and climatology. The data are for varying numbers of years, ranging from 19 years (1958–1976) for Key West, Florida, to 35 years (1943–1977) for Brownsville, Texas, so that we should prefer a common basis for graphical comparison of these univariate samples. `windspeed.dta` is provided with the media for this issue of the *Stata Journal*. The dataset includes two variables, `windspeed` and `place`. The ordering of the places is by mean maximum windspeed.

Researchers accustomed to such data tend to reach first for quantile plots. The official command `quantile` is limited to one batch of data at a time. While more versatile user-written alternatives are available (Cox 2005a, 2010), the spirit of this particular column is that you can work out code for yourself using basic commands.

Given an ordered sample of size n for variable y, $y_{(1)} \leq y_{(2)} \ldots y_{(n-1)} \leq y_{(n)}$, the usual ordinate and abscissa for a quantile plot are $y_{(i)}$ and $(i - a)/(n - 2a + 1)$, respectively. The abscissa for some choice of a is, in effect, an empirical cumulative probability and is often called a *plotting position*. The naïve choice i/n for plotting position would imply probabilities $1/n$ and 1 at the ends of the data, while $(i - 1)/n$ would just reverse the problem. Either choice would be awkward, implying that no value can be more extreme than those observed and because theoretical quantiles are often not defined for probabilities 0 or 1. We need, therefore, a slightly more complicated method. The choice of a is the subject of a small but contentious literature, to which Thas (2010) is one entry point. Given a choice, we can implement it for ourselves in Stata. Below we use $a = 0$—that is, $i/(n + 1)$, as is common in statistics of extremes.

To calculate plotting positions, it is convenient, if not outstandingly efficient, to use `egen` functions. These functions take care of sorting issues, handling of any missing values, and separate calculations for separate groups:

```
. use windspeed
. egen rank = rank(windspeed), by(place) unique
. egen count = count(windspeed), by(place)
. generate pp = rank/(count + 1)
. label variable pp "fraction of data"
```

Figure 1 shows quantile plots, with a separate panel for each place.

```
. scatter windspeed pp, by(place) yla(, ang(h)) xla(0(.25)1)
```

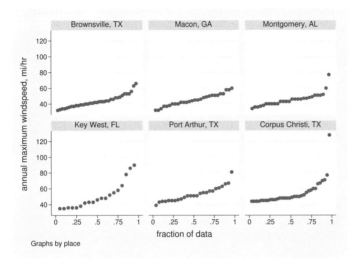

Figure 1: Quantile plots of annual maximum windspeed data for six places in the southeastern United States

The usual starting point with such data is the fitting of Gumbel distributions, with a distribution function for location parameter (mode) ξ and scale parameter α of

$$F(y) = \exp[-\exp\{-(y - \xi)/\alpha\}]$$

defined over the real line. Gumbel distributions are named for Emil Julius Gumbel (1891–1966), who did much to systematize knowledge of the statistics of extremes and wrote the first extended monograph on the subject (Gumbel 1958). For more detail on his scientific and political career, see Freudenthal (1967), Hertz (2001), and Brenner (2001).

For context, note that the mean of a Gumbel distribution is $\xi + \alpha\gamma$ and the standard deviation is $\alpha\pi/\sqrt{6}$. Here $\gamma \approx 0.57721+$ is Euler's constant (in Stata, -digamma(1)) and $\pi \approx 3.14159$ (in Stata, _pi or c(pi)) is the even better known constant. Without venturing into numerical fits, the distribution function can easily be inverted, giving

$$y(F) = \xi - \alpha \ln(-\ln F)$$

so that a plot of y against $-\ln(-\ln F)$ should be approximately linear with intercept ξ and slope α if y is drawn from a Gumbel distribution. The quantity $-\ln(-\ln F)$ is thus often called a Gumbel reduced variate, "reduced" implying unit-free and dimensionless. The resulting plot is a Gumbel plot. For another example and literature references, see Cox (2007a).

Because the plotting position variable pp has already been calculated separately for each place, we can apply functions as just stated algebraically:

```
. generate gumbel = -ln(-ln(pp))
. label variable gumbel "Gumbel reduced variate"
```

There are two obvious versions of the corresponding graph. Figure 2 separates out different places into different panels:

```
. scatter windspeed gumbel, by(place) yla(, ang(h))
```

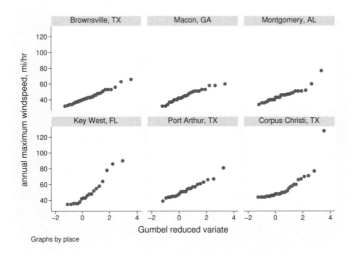

Figure 2: Gumbel plots of annual maximum windspeed data for six places in the southeastern United States, one panel for each place

Figure 2 is clearly ideal for considering individual places. How easy does it make comparison, however? Conversely, figure 3 separates places by using different point symbols (Cox 2005b). The separate command (see [D] **separate**) makes this step easier but is not essential because a series of if qualifications could produce the same result.

```
. separate windspeed, by(place) veryshortlabel
. scatter windspeed? gumbel, ytitle("`: var label windspeed'") yla(, ang(h))
> legend(pos(11) ring(0) order(6 5 4 3 2 1) col(1))
```

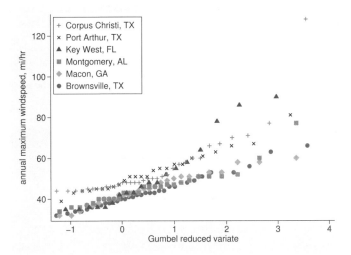

Figure 3: Gumbel plots of annual maximum windspeed data for six places in the southeastern United States, places being superimposed

I initially ordered the places lowest mean first. Now it becomes evident that the reverse order is needed here. More importantly, graphs like figure 3 appear frequently in books and journals. But how effective are they? Readers will find it easy to understand the principle: given detailed study of the legend, they could study the graph carefully to learn more about contrasts. But will they be encouraged to do that by the design? And how easy would that be? It is a characteristic of these data, like many others, that the groups overlap to some extent. With this design, however, that inevitably implies that some groups are partially obscured on the graph by others.

The resulting dilemma is clear. Figure 2 and figure 3 have corresponding advantages and limitations. Moreover, the example should strike many readers as modest if not minute in size, with just 6 groups and sample sizes between 19 and 35. The problem with more data can be much more serious.

A suggested compromise is this: Show each group separately, but with the rest of the data shown as a backdrop. Figure 4 is the result. Now, as is natural in many ways, the other data provide context for each subset.

```
. quietly forvalues i = 1/6 {
>     scatter windspeed gumbel if place != `i', ms(Oh) mcolor(gs12)
>     || scatter windspeed gumbel if place == `i', ms(D) mcolor(gs1)
>     yla(, ang(h)) yti("") xti("") legend(off)
>     subtitle("`: label (place) `i''", box fcolor(gs13) bexpand size(medium))
>     name(g`i', replace)
> }
```

```
. graph combine g1 g2 g3 g4 g5 g6, imargin(small)
> l2ti("`: var label windspeed´") b2ti("`: var label gumbel´")
```

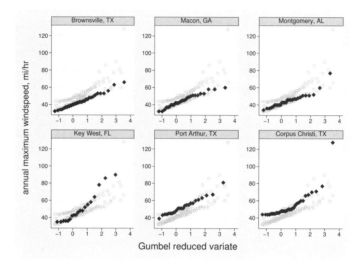

Figure 4: Gumbel plots of annual maximum windspeed data for six places in the southeastern United States. Data for the five other places are shown as a backdrop to the data for each place.

The code is more complicated than for previous graphs, but the logic is still straightforward. The next subsection gives a commentary for those who would like to see matters discussed in greater detail. The details of what you see printed depend on having previously set the *Stata Journal* graph scheme by typing

```
. set scheme sj
```

2.2   Comments on code

```
quietly forvalues i = 1/6 {
```

1. We loop using `forvalues` over the distinct groups of a categorical variable (`place`).
   In this case, we know in advance that there are 6 groups, numbered 1–6. In more
   general situations, we might want to automate the looping. Various techniques
   exist for doing that. One method is to use `levelsof` (see [P] **levelsof**) to produce
   a list of the distinct groups, followed by a call to `foreach`. Another, perhaps
   easier, method is to use the `group()` function of `egen` (see [D] **egen**) to define a
   grouping variable with positive integer values (Cox 2007b).

```
        scatter windspeed gumbel if place != `i´, ms(Oh) mcolor(gs12)
```

2. We first lay down the rest of the data as a backdrop. So, for example, the first time around the loop, when 'i' evaluates to 1, we look for `place` not equal to 1. The data for that complementary subset are shown lightly. The suggestion here is that open circles `Oh` and a light color `gs12` are suitably muted.

   It is tempting just to plot all the data as substrate, on the grounds that we will just plot over each subset being emphasized. In principle, that is correct; in practice it is possible for small parts of the symbols in question to be visible even though they lie underneath the symbols to be plotted. So do use the `!=` constraint.

```
     || scatter windspeed gumbel if place == `i´, ms(D) mcolor(gs1)
```

3. The subset being emphasized is now plotted directly on top. More prominent symbols and colors are needed. (In other problems, line patterns, widths, and colors will be the elements to adjust.)

```
        yla(, ang(h))  yti("") xti("") legend(off)
```

4. About the options specified, note first that `yla(, ang(h))` is a personal choice, although the underlying logic that text is more readable horizontally is a point on which many will agree. More particular to the key themes here are suppression of `ytitle()`, `xtitle()`, and `legend()`. In this problem, the `ytitle()` and `xtitle()` would be the same on all six graphs, which is unnecessary. We will see in a moment how to put just one title on both the left and bottom sides of the overall graph. The appearance of a legend would be triggered by the double plotting of `windspeed`. A legend would not help, so we suppress it, too.

```
     subtitle("`: label (place) `i´´", box fcolor(gs13) bexpand size(medium))
```

5. Evidently, each graph needs some explanatory text. `place` is a numeric variable with value labels attached, so we use an extended macro function (see [P] **macro**) to look up the label concerned as we go around the loop. If no label were attached to the numeric value in question, the value itself would be shown instead. This approach does not extend to string variables, except that there is an easy work-around: just map the string variable to a numeric variable with value labels first, using `encode` (see [D] **encode**) or the `egen` function `group()` that was mentioned earlier.

   In the rendering of the text, the extra options `box`, `fcolor(gs13)`, `bexpand`, and `size(medium)` are doubly optional. In this case, they come from peeking at graphs produced with `by()` in the Graph Editor and so producing a similar overall style.

```
        name(g`i´, replace)
```

6. It is essential that we save each graph for later combining, which is the next step. There is a choice between using `name()` and using `saving()`. A side effect of using `name()` is that each resulting graph remains open in a separate Graph window, so that it can be checked. Although we have not previously used any of these graph

names, writing "`, replace`" will let you revise this code a little more easily—that is, assuming that you do not write perfect code the first time and every time.

```
graph combine g1 g2 g3 g4 g5 g6, imargin(small)
l2ti("`: var label windspeed´") b2ti("`: var label gumbel´")
```

7. `graph combine` is used to put the graphs together. In this case with six graphs, the default of two rows and three columns looks fine. We add titles to the combined graph using `l2title()` and `b2title()`. The option names denote titles on the `left` and `bottom` of the graph (and may evoke nostalgia among longtime Stata users for the graphics syntax used before Stata 8). Notice further how we generalize a step beyond wiring in the particular variable labels: the extended macro option calls ensure that Stata looks up the current variable labels, so that the same code can be used even if we change the variable labels. (More general code yet would protect against the possibility that no variable labels have been assigned.)

3   Intermezzo: Advice from Edward Tufte

Make all visual distinctions as subtle as possible, but still clear and effective (Tufte 1997, 73).

Minimal contrasts of the secondary elements (figure) relative to the negative space (ground) will tend to produce a visual hierarchy with layers of inactive background, calm secondary structure, and notable content. Conversely, when *everything* is emphasized, *nothing* is emphasized; the design will often be noisy, cluttered, and informationally flat (Tufte 1997, 74).

4   A bivariate example: Cirque lengths, widths, and grades

For a bivariate example, we examine some data similar to a dataset used in Cox (2005b). The data (Evans and Cox 1995) refer to the lengths and widths of cirques in the Lake District, England. `cumbrian_cirques.dta` is provided with the media for this issue of the *Stata Journal*. Cirques are armchair-shaped hollows formerly occupied by glaciers. Length and width are basic quantitative measures of their size. Logarithmic scales are standard for such data. Here we also bring in grade, a judgment-based variable of how well developed each feature is on a five-point ordered scale from classic to poor.

Figure 5 is a standard scatterplot. Because we will have five scatterplots for each grade, we might as well use the `total` suboption to add a panel for all the data combined. An ad hoc refinement is to insist on an extra space in the axis label at 2000 meters to prevent the last digit from being elided in the combined display.

```
. use cumbrian_cirques, clear
. scatter width length, by(grade, total) xsc(log) ysc(log) ms(Oh)
> xla(200 500 1000 2000 "2000 ") yla(200 500 1000 2000)
```

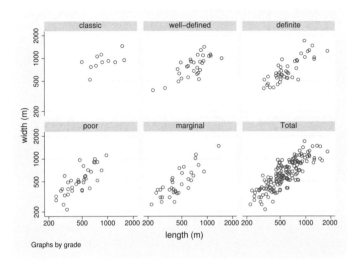

Figure 5: Scatterplots of cirque width and length by grade for the English Lake District

In the compromise design, most of the small code tricks are the same, but as with figure 5 we add a further display of all the data. So that you have a comparison of style with earlier graphs, we will leave the subtitle area unboxed. Figure 6 is the result.

```
. forvalues i = 1/5 {
>     scatter width length if grade != `i´, xsc(log) ysc(log) ms(Oh) mcolor(gs12)
>     xla(200 500 1000 2000 "2000 ") yla(200 500 1000 2000)
>     || scatter width length if grade == `i´, xsc(log) ysc(log) ms(D) mcolor(gs1)
>     yla(, ang(h)) yti("") xti("") legend(off)
>     subtitle("`: label (grade) `i´´", size(medium)) name(g`i´, replace)
> }
. scatter width length, xsc(log) ysc(log) ms(D) mcolor(gs1)
> xla(200 500 1000 2000 "2000 ") yla(200 500 1000 2000)
> yla(, ang(h)) yti("") xti("") legend(off)
> subtitle("all cirques", size(medium)) name(g6, replace)

. graph combine g1 g2 g3 g4 g5 g6, imargin(small)
> l2ti("`: var label width´") b2ti("`: var label length´")
```

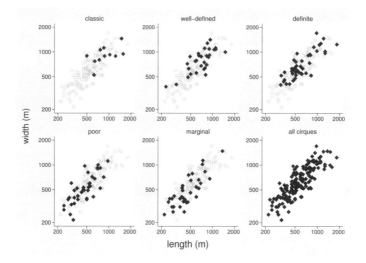

Figure 6: Scatterplots of cirque width and length by grade for the English Lake District. Data for the four other grades are shown as a backdrop to the data for each grade.

5 Conclusions

The conclusions lie with you, the reader. This column has a flavor of experiment. Do you think that the compromise design—which might be called a subset and substrate design—has any advantages over alternatives for the examples here? Do you think that you can use the ideas here to improve your own comparative displays? Many graph types lie open for exploration, including especially attempts to more easily and more effectively see fine structure in spaghetti plots.

6 References

Brenner, A. D. 2001. *Emil J. Gumbel: Weimar German Pacifist and Professor*. Boston: Brill.

Cleveland, W. S. 1993. *Visualizing Data*. Summit, NJ: Hobart.

————. 1994. *The Elements of Graphing Data*. Rev. ed. Summit, NJ: Hobart.

Cox, N. J. 2005a. Speaking Stata: The protean quantile plot. *Stata Journal* 5: 442–460.

————. 2005b. Stata tip 27: Classifying data points on scatter plots. *Stata Journal* 5: 604–606.

————. 2007a. Stata tip 47: Quantile–quantile plots without programming. *Stata Journal* 7: 275–279.

————. 2007b. Stata tip 52: Generating composite categorical variables. *Stata Journal* 7: 582–583.

————. 2009. Stata tip 78: Going gray gracefully: Highlighting subsets and downplaying substrates. *Stata Journal* 9: 499–503.

————. 2010. Software update: gr42_5: Quantile plots, generalized. *Stata Journal* 10: 691–692.

Evans, I. S., and N. J. Cox. 1995. The form of glacial cirques in the English Lake District, Cumbria. *Zeitschrift für Geomorphologie* 39: 175–202.

Freudenthal, A. M. 1967. Emil J. Gumbel. *American Statistician* 21(1): 41.

Gumbel, E. J. 1958. *Statistics of Extremes.* New York: Columbia University Press.

Hertz, S. 2001. Emil Julius Gumbel. In *Statisticians of the Centuries*, ed. C. C. Heyde and E. Seneta, 406–410. New York: Springer.

Hosking, J. R. M., and J. R. Wallis. 1997. *Regional Frequency Analysis: An Approach Based on L-Moments.* Cambridge: Cambridge University Press.

Robbins, N. B. 2010. Trellis display. *Wiley Interdisciplinary Reviews: Computational Statistics* 2: 600–605.

Simiu, E., M. J. Changery, and J. J. Filliben. 1979. Extreme wind speeds at 129 stations in the contiguous United States. Building Science Series 118, National Bureau of Standards.

Thas, O. 2010. *Comparing Distributions.* New York: Springer.

Tufte, E. R. 1990. *Envisioning Information.* Cheshire, CT: Graphics Press.

————. 1997. *Visual Explanations: Images and Quantities, Evidence and Narrative.* Cheshire, CT: Graphics Press.

————. 2001. *The Visual Display of Quantitative Information.* 2nd ed. Cheshire, CT: Graphics Press.

————. 2006. *Beautiful Evidence.* Cheshire, CT: Graphics Press.

About the author

Nicholas Cox is a statistically minded geographer at Durham University. He contributes talks, postings, FAQs, and programs to the Stata user community. He has also coauthored 15 commands in official Stata. He wrote several inserts in the *Stata Technical Bulletin* and is an editor of the *Stata Journal*.

The Stata Journal (2012)
12, Number 2, pp. 332–341

Speaking Stata: Transforming the time axis

Nicholas J. Cox
Department of Geography
Durham University
Durham City, UK
n.j.cox@durham.ac.uk

Abstract. The time variable is most commonly plotted precisely as recorded in graphs showing change over time. However, if the most interesting part of the graph is very crowded, then transforming the time axis to give that part more space is worth consideration. In this column, I discuss logarithmic scales, square root scales, and scale breaks as possible solutions.

Keywords: gr0052, axis scale, graphics, logarithm, scale break, square root, time, transformation

1 Introduction

Using transformed scales on graphs is a technique familiar to data analysts. Most commonly in Stata, axis scales on a `twoway` graph may be logarithmic. The options `yscale(log)` and `xscale(log)` provide wired-in support. In principle, other transformed scales are hardly more difficult: just calculate the transformation before graphing and, ideally, ensure that the axis labels remain easy to interpret. For example, a logit scale can be useful for proportions within $(0, 1)$ or percentages within $(0, 100)$. For that example and more general comments, see Cox (2008).

In the case of graphs plotting change over time, the time axis usually shows the time variable unchanged. In most sciences, time is plotted on the horizontal or x axis. In the earth sciences, time is sometimes plotted on the vertical or y axis, essentially because older materials are often found at greater depths.

In either case, however, it is common to use a transformed scale for whatever response is plotted versus time. For example, with a logarithmic response scale, periods of exponential growth or decline would be shown by straight line segments.

In this column, I focus on something more unusual: transforming the time axis. The argument is in the first instance pragmatic. Sometimes data span an extended time range, and the amount of information or the pace of change is far greater at one end of the time range. If so, expanding the crowded part of the graph and shrinking the other part of the graph may help readers visualize the detail of change.

If data come from cosmology or particle physics, blowing up the early part of a time range might be a good idea. If data come from the social sciences, especially economic or demographic history, blowing up the later part might be called for. In many examples, very recent growth has been explosive, but the precise details remain of interest. The same recommendation applies, but for different reasons, to many datasets from the earth sciences. Here information from earlier periods tends to be much sparser and indeed in most instances has long since been buried, blasted, or eroded away.

2   Logarithmic scale?

A logarithmic scale for time would show earlier times—all later than some origin, itself conventionally zero—more expansively. `xscale(log)` should suffice, assuming again that time is shown on the x axis. Conversely, a logarithmic scale for times, all those shown being *before* some origin, would do the same for later times. In the second case, you would need to calculate the logarithm of (origin $-$ time) and devise intelligible labels.

However, a major limitation in both cases is that the origin is not plottable because the logarithm of zero is indeterminate. A fix for this is to ensure that the origin is not a data point but is beyond the range of the data. Otherwise put, if time 0 is a data point, the origin must be offset. Karsten (1923, chap. 41) called plotting in terms of logarithm of (offset $+$ latest $-$ time) "retrospective logarithmic projection" and gave economic examples. This common need for an offset is at best awkward and at worst leads to arbitrary choices that are difficult to justify or explain.

It remains true that thinking logarithmically is highly desirable for thinking broadly about the ranges of time scales (and sizes, speeds, and numerosity) to be found in science. See, for example, the engaging essay "Life on log time" in Glashow (1991) or the more extended discussions in Schneider (2009).

Although using a logarithmic scale for either variable or both variables is likely to strike readers as an elementary and obvious technique, its history is in several ways surprisingly short (Funkhouser 1937, 359–361). Lalanne (1846) introduced double logarithmic scales for calculation problems: Hankins (1999, 71) reproduces his graph. In economics, Jevons (1863, 1884) plotted data for responses on a logarithmic scale versus time; for discussion, see Morgan (1990), Klein (1997), and Stigler (1999). Nevertheless, adoption of the idea was slow, and it was even resisted as obscuring the data. Expository articles urging the merits of the technique, often called semilogarithmic or ratio charts, were still appearing many years after its introduction (Fisher 1917; Field 1917; Griffin and Bowden 1963; Burke 1964).

3   Square root scale?

In practice in data analysis, however, logarithmic transformations are likely to be too strong unless change on very different time scales is being considered. A weaker transformation that can work well is the square root, as recently suggested by Wills (2012, 171–172). The square root also has the simple advantage that the root of 0 is also 0 and is thus quite plottable.

The earliest example of a square root scale known to me is by Moseley (1914) in a classic physics problem that led to predictions of as yet undiscovered elements. Heilbron (1974, 100) reproduces his graph. Karsten (1923, 485) plotted the square root of response versus time in an example in which a logarithmic scale was clearly too severe. Fisher and Mather (1943) plotted frequency counts on a square root scale. Their idea was taken up in much more detail by Mosteller and Tukey (1949), and Fisher (1944, 1950) added corresponding material to the ninth and eleventh editions of his text *Statistical Methods for Research Workers*.

There are contexts ranging from physics to finance in which the square root of time appears naturally, or at least mathematically, in discussions, but we leave that to one side. In the example to follow, the justification for using square roots is, once more, just pragmatic.

Consider some data from Haywood (2011) on world population over the last 8,000 years, available with the media for this journal issue. As is customary, the cautions and caveats associated with such estimates should be flagged. Comprehensive censuses are relatively recent historically, and even long national experience in census-taking is no guarantee of population counts that all experts regard as accurate. Over most of human history, we have only a variety of wild guesses, and competent workers in the field often only agree if they are using each other's estimates. Let us look first at some standard plots.

```
. use haywood

. twoway connected pop year, msymbol(Oh)

. twoway connected pop year, msymbol(Oh) yscale(log)
> ylabel(20 50 200 500 2000 5000)
```

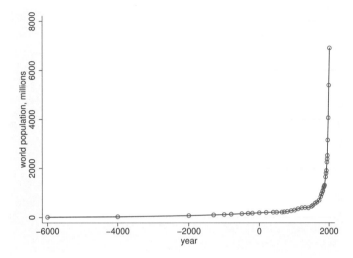

Figure 1: World population over the last 8,000 years (data from Haywood [2011])

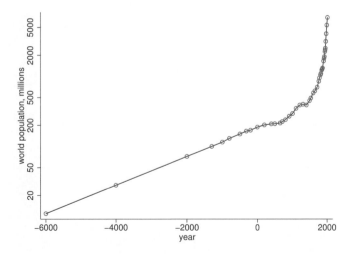

Figure 2: World population over the last 8,000 years; the vertical scale is logarithmic, so periods showing exponential change (constant growth rate) would plot as linear segments (data from Haywood [2011])

Figure 1 draws attention to the explosive nature of population growth over the last few centuries. It has a simple and powerful message, but detail is hard to discern. Figure 2 adds a logarithmic scale and shows a main pattern of accelerating growth rates with strong hints of other fluctuations, reflecting extended periods of higher death rates. For both graphs, using `twoway connected`, a hybrid of `scatter` and `line`, has a natural purpose of emphasizing the precise dates of data points. With more data or data regularly spaced in time, using `line` is likely to be a better choice. That detail aside, figure 1 and figure 2 are standard graphs that are both useful, but neither deals with the problem of crowding of data for later times.

Using a square root scale for time before the present requires two steps before we can create a graph. First, calculate the new time scale.

```
. generate time = sqrt(2010 - year)
```

The latest year for data is 2010. Clearly, other datasets would lead to different constants.

Second, produce better axis labels. The key here is that any text you like can be shown as text in axis labels. In practice, we need to work out what labels we want to see and then where they would be put in terms of the new variable, here `time`. This can take some experimentation, but it is best done by looping over the labels we want and working out the option call that is needed.

```
. foreach y of num -6000(2000)2000 500 1000 1500 {
.        local call `call' `=sqrt(2010 - `y')' "`y'"
. }
```

Here the result is put into a local macro. Every time around the loop, we calculate the square root of 2010 minus some year and add that value and the text to be shown on the time axis—namely, the year itself—as extra content in the local macro. So the first time around, the year is -6000. We use Stata to calculate `sqrt(2000 - (-6000))` on the fly and insert the result in the macro. The special syntax '`=expression`' instructs Stata to calculate the result of the *expression* and use that result immediately.

If you do the calculation for -6000 yourself with `display`, you will find that the result is given as 89.442719. On-the-fly calculation gives more decimal places, far more than are really needed for graphical purposes, but no harm is done thereby. The calculation

```
89.44271909999159 "-6000"
```

is added to the macro, which is created empty. The next time around, the macro will contain

```
89.44271909999159 "-6000" 77.45966692414834 "-4000"
```

and so on.

If you do this yourself, make sure to blank out the local macro by typing

```
. local call
```

before revising any earlier guess. Otherwise, earlier insertions in the macro will just be carried forward.

Now the graph shown as figure 3 is possible. Reversing the scale is needed because the time variable increases going backward, at least if you want to show the direction of time in the usual manner.

```
. twoway connected pop time, msymbol(Oh) xscale(reverse) xlabel(`call')
> xtitle(year)
```

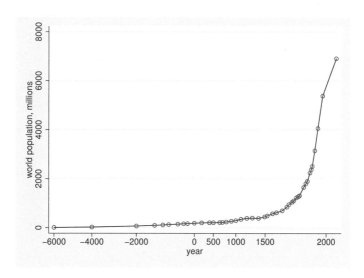

Figure 3: World population over the last 8,000 years; the horizontal axis represents time before 2010 on a square root scale (data from Haywood [2011])

You can see the advantage of automating the production of the option call: you would not prefer to do all the little calculations one by one with a flurry of copy-and-paste to transfer results to the graph command.

Nothing stops the use of `yscale(log)` too, although the standard result that exponential change shows as linear segments no longer holds. My own preference is to stop at transforming the time scale.

As usual, smaller graphical details remain open to choice. The axis title of "year" might seem redundant. A method other than negative dates might be preferred for showing years BCE. The *y*-axis labels could be rotated to horizontal.

4   Scale break?

Another popular solution to the problem of crowding for early or later times is to use
a scale break so that the time series is divided into two or more panels with different
scales. Scale breaks themselves divide statistically minded people. Some argue for them
as a simple practical solution, so long as it is made explicit, and some argue against them
as awkward, artificial, and usually unnecessary, because transformation is typically a
better solution. Good discussions are given by Cleveland (1994) and Wilkinson (2005).
For more discussion of scale breaks in Stata graphics, see Cox and Merryman (2006).

Stata does not support scale breaks directly, perhaps mostly because a graph with a
scale break is not easily made consistent with the general philosophy of allowing graphs
to be superimposed or combined with a common axis. But you can devise your own
scale breaks by producing separate graphs before combining them again.

```
. twoway connected pop y if year < 0, msymbol(Oh) yscale(r(10 7000) log)
> ylabel(10 20 50 100 200 500 1000 2000 5000, angle(h))
> saving(part1, replace) xlabel(-6000(1000)0) xtitle("")
. twoway connected pop y if year > 0, ms(Oh) ysc(r(10 7000) log off)
> yla(10 20 50 100 200 500 1000 2000 5000, ang(h))
> saving(part2, replace) xla(250(250)2000) xtitle("")
. graph combine "part1" "part2", imargin(small) b1title(year, size(small))
```

The result is shown in figure 4. It is likely to qualify as many users' best bet for this
kind of problem, if only because it is easier to explain than other solutions.

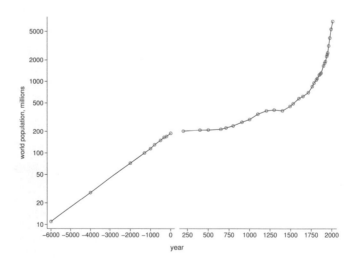

Figure 4: World population over the last 8,000 years; scale break at change from BCE
to AD (data from Haywood [2011])

Let us focus on some of the small details in these commands that might otherwise seem puzzling.

1. I tend to include the `replace` suboption of `saving()` even the first time around, because in practice I need several iterations before all my choices seem right. Nevertheless, do check that you are not losing a valuable graph from some previous project.

2. `graph combine` has a `ycommon` option. But with `yscale(log)` too, I get better results by being explicit in `yscale()` in both graph commands that the *y* axes have the same limits.

3. Why spell out the `ylabel()` call for the second graph if the labels are omitted by `yscale(off)`? The reason is that the label locations determine horizontal grid lines. (Naturally, if you did not want those grid lines, you should omit them.)

4. The title "year" should certainly not be given twice, so we blank it out in the individual graph commands and put it back (once) in the `graph combine` command.

5. The first panel ends up a bit smaller in this example as a side-effect of omitting the *y* axis and its labels from the second panel. That seems about right, but the relative sizes of the panels could be further controlled by using the `fxsize()` option with the initial graph commands. Cox and Merryman (2006) give an example.

In principle, two or more scale breaks are possible, giving three or more panels. You just need to produce the individual graph panels as separate graphs and then combine them. But the trade-offs are clear. Each panel must be big enough to make clear what the different scales are, but without overloading the time axis with many different labels or ticks. For that and other reasons, even a graph with two scale breaks and three panels is likely to be too complicated for most purposes.

5   Conclusions

For showing change over time graphically, transformation almost always means transforming the response. Logarithmic scales are by far the most widely used example and need little publicity. In this column, I focused on a less common but intriguing possibility: transforming time. Although logarithmic scales have been used on occasion for time, they can be awkward at best. In contrast, a square root scale for time is easy to implement and can work well. Using a scale break to divide a graph into separate panels is also a possibility.

As always, when users have choices, they also have to make decisions on what to do. The question is likely to depend on specific circumstances of dataset, analysis, and audience, as well as researchers' general attitudes. In choosing any of the graphs here for a presentation, what the audience would find easiest to understand—or in some cases, what they would find most stimulating to consider—would be the first question to consider.

6 Acknowledgment

Stephen M. Stigler provided invaluable details for references to successive editions of R. A. Fisher's text.

7 References

Burke, T. 1964. Semi-logarithmic graphs in geography: A pertinent addendum. *Professional Geographer* 16(1): 19–21.

Cleveland, W. S. 1994. *The Elements of Graphing Data*. Rev. ed. Summit, NJ: Hobart.

Cox, N. J. 2008. Stata tip 59: Plotting on any transformed scale. *Stata Journal* 8: 142–145.

Cox, N. J., and S. Merryman. 2006. FAQ: How can I show scale breaks on graphs? http://www.stata.com/support/faqs/graphics/scbreak.html.

Field, J. A. 1917. Some advantages of the logarithmic scale in statistical diagrams. *Journal of Political Economy* 25: 805–841.

Fisher, I. 1917. The "ratio" chart for plotting statistics. *Publications of the American Statistical Association* 15: 577–601.

Fisher, R. A. 1944. *Statistical Methods for Research Workers*. 9th ed. Edinburgh: Oliver & Boyd.

———. 1950. *Statistical Methods for Research Workers*. 11th ed. Edinburgh: Oliver & Boyd.

Fisher, R. A., and K. Mather. 1943. The inheritance of style length in *Lythrum salicaria*. *Annals of Eugenics* 12: 1–23.

Funkhouser, H. G. 1937. Historical development of the graphical representation of statistical data. *Osiris* 3: 269–404.

Glashow, S. L. 1991. *The Charm of Physics*. New York: American Institute of Physics.

Griffin, D. W., and L. W. Bowden. 1963. Semi-logarithmic graphs in geography. *Professional Geographer* 15(5): 19–23.

Hankins, T. L. 1999. Blood, dirt, and nomograms: A particular history of graphs. *Isis* 90: 50–80.

Haywood, J. 2011. *The New Atlas of World History: Global Events at a Glance*. London: Thames & Hudson.

Heilbron, J. L. 1974. *H. G. J. Moseley: The Life and Letters of an English Physicist, 1887–1915*. Berkeley, CA: University of California Press.

Jevons, W. S. 1863. *A Serious Fall in the Value of Gold Ascertained, and Its Social Effects Set Forth.* London: Edward Stanford.

———. 1884. *Investigations in Currency and Finance.* London: Macmillan.

Karsten, K. G. 1923. *Charts and Graphs: An Introduction to Graphic Methods in the Control and Analysis of Statistics.* New York: Prentice Hall.

Klein, J. L. 1997. *Statistical Visions in Time: A History of Time Series Analysis 1662–1938.* Cambridge: Cambridge University Press.

Lalanne, L. 1846. Mémoire sur les tables graphiques et sur la géométrie anamorphique appliquée à diverses questions qui se rattachent à l'art de l'ingénieur. *Annales des Ponts et Chaussées* 11: 1–69.

Morgan, M. S. 1990. *The History of Econometric Ideas.* Cambridge: Cambridge University Press.

Moseley, H. G. J. 1914. The high-frequency spectra of the elements. Part II. *Philosophical Magazine*, 6th ser., 27: 703–713.

Mosteller, F., and J. W. Tukey. 1949. The uses and usefulness of binomial probability paper. *Journal of the American Statistical Association* 44: 174–212.

Schneider, D. C. 2009. *Quantitative Ecology.* 2nd ed. London: Academic Press.

Stigler, S. M. 1999. *Statistics on the Table: The History of Statistical Concepts and Methods.* Cambridge, MA: Harvard University Press.

Wilkinson, L. 2005. *The Grammar of Graphics.* 2nd ed. New York: Springer.

Wills, G. 2012. *Visualizing Time: Designing Graphical Representations for Statistical Data.* New York: Springer.

About the author

Nicholas Cox is a statistically minded geographer at Durham University. He contributes talks, postings, FAQs, and programs to the Stata user community. He has also coauthored 15 commands in official Stata. He wrote several inserts in the *Stata Technical Bulletin* and is an editor of the *Stata Journal*.

The Stata Journal (2012)
12, Number 3, pp. 549–561

Speaking Stata: Axis practice, or what goes where on a graph

Nicholas J. Cox
Department of Geography
Durham University
Durham City, UK
n.j.cox@durham.ac.uk

Abstract. Conventions about what information goes on each axis of a two-way plot are precisely that, conventions. This column discusses—historically, syntactically, and by example—the idea that flouting convention in various ways can lead to small but useful improvements in graph display. Putting y-axis information on the right or on the top, or putting x-axis information on the top, often is useful. The most substantial examples are for multiple quantile plots, for which the new command `multqplot` is offered, and table-like graphs, which are made even more table-like by mimicking column headers.

Keywords: gr0053, multqplot, qplot, tabplot, axes, coordinates, quantile plots, two-way bar charts

1   Introduction

What goes where on a rectangular graph (which means most graphs)? This question is easy to answer, if only with tongue in cheek. Plot what you think of as response or outcome on the vertical or y axis unless you have good reason to do otherwise. The "otherwise", for example, can include cases where convention indicates that depth or height should be plotted on the vertical axis. Even if that has been decided, there remains the smaller but still interesting question of where axis titles, labels, and ticks for such variables are best placed.

Such a specific question raises a broader issue: How far is there a binding logic to graphical choices, and how far are we free to follow conventions or to flout them according to taste or circumstance?

The key questions are made concrete by focusing on graphs that are implemented in terms of Stata commands such as `twoway scatter`, `twoway line`, or `twoway connected`. For such graphs, Stata's defaults will be familiar to you: y-axis information goes on the left and to the left of the axis line, and x-axis information goes on the bottom and beneath the axis line. This is a widespread convention in science and mathematics that presumably arises because many graphs show quantities that are 0 or positive and also show axes that are $y = 0$ and $x = 0$. Thus putting extra explanatory information outside the data region and next to the axes seems natural.

In this column, I will discuss situations in which subverting this convention makes sense and how to do that in Stata.

2   A little history

Graphical choices should be defended as matters of logic when there is an underlying logic; otherwise, they may just be explained as matters of taste, convenience, or convention. Even so, a look at the history of axes and coordinates is instructive, and perhaps even surprising. Sources that are especially informative or entertaining include Cajori (1929), Boyer (1956), Bochner (1966), Kline (1972), and Stillwell (2010).

The idea of coordinates was latent in Greek mathematics, in the work of Nicole Oresme (c. 1320–1382), and indeed throughout the practices of surveying, cartography, navigation, and astronomy. The key breakthrough of seeing that geometry and algebra could be united as coordinate or analytic geometry to the benefit of both is generally attributed to Pierre de Fermat (1601–1665) and René Descartes (1596–1650).

Descartes was responsible for the present convention of using letters such as x, y, and z for coordinates and receives enduring credit through the still widespread term *Cartesian coordinates*. Nevertheless, the work of Fermat and Descartes was still some distance from what is now standard. According to Bochner (1966, 40), "coordinate systems in the full sense of the notion only *begin* to occur in the work of Descartes himself". The explicit use of both x and y axes, or a z axis when needed, and the idea that either x or y might be negative became more evident in the later 17th and 18th centuries. Histories of mathematics differ on who introduced precisely what and when, but outstanding mathematicians such as John Wallis (1616–1703), Isaac Newton (1643–1727), Gottfried Wilhelm Leibniz (1646–1716), who introduced the term *coordinates*, and Leonhard Euler (1707–1783) all contributed.

Over the last two hundred years, practices within what we might now call statistical graphics have been far from consistent. Strikingly, William Playfair (1759–1823) in his *Commercial and Political Atlas* (1801a) shows all his time series with y-axis labels on the right, corresponding to the most recent data, and with an explanation of what is being shown in a boxed title within the data region. He also uses y-axis labels on the right in many of his other graphs, for example, in his *Statistical Breviary* (1801b). Both 1801 works were reprinted together in Playfair (2005).

A century or so later, Brinton's *Graphic Methods for Presenting Facts* (1914) gave a cross-section of current practice and much advice on good and bad technique. (His admonitions against what are now usually called pie charts and against a third dimension introduced in graphs for artistic effect are a foretaste of often repeated, and often ignored, criticisms from most more recent writers on statistical graphics.) On page 361, Brinton proposed 2 of 25 rules offered "simply as suggestions":

> Figures for the horizontal scale should always be placed at the bottom of the chart. If needed, a scale may be placed at the top also.

> Figures for the vertical scale should always be placed at the left of the chart. If needed, a scale may be placed at the right also.

However, such principles are unlikely to persuade those otherwise inclined. Specifically, several examples in Brinton's eclectic collection violate both rules. More generally, what force does a rule have if it is merely a suggestion and there is no reason given for why it is to be the standard?

Nevertheless, most statistical graphics over the last century has followed Brinton's advice, so much so that many texts do not even spell out that these are the conventions. Court (1987, 1642) does precisely that: "On rectangular diagrams, primary scales at bottom and left should increase consistently upward and toward the right, away from the axis intersection; secondary scales at top and right may give equivalent units (metric on primary and English on secondary, or frequency at left and wavelength at right) or may apply to a second set of data." This is clear enough, but Court nowhere explains why it would be illogical to do otherwise.

3   Flouting conventions

To cut to the chase, I will give examples of good reasons to flout convention.

1. When you plot time series, both theory and practice often imply most interest in most recent values. In addition, if time is a calendar date, 0 is rarely shown in any case (and if it is shown, it is usually in conjunction with dates BCE, as in my previous column [Cox 2012b]). So putting y-axis labels on the right puts them where they will be easiest to use. This was presumably Playfair's logic, and it has appealed to others since. Klass (2012) is a friendly recent text with several social science examples. In fact, time-series graphs with variable information at the top and axis labels on the right are commonplaces of financial journalism; for example, have a look at issues of the *Economist*, *Financial Times*, or the *Wall Street Journal*.

2. When you plot several graphs together, pressure on space is much increased. Whenever the x axis shows a common variable, the top of the graph can be a good place for titles explaining the y variable, which traditionally are placed on the y axis. This also satisfies the simple but often ignored idea that text is easiest to read if words run from left to right.

3. Table-like graphs will not suffer from more use of table-like conventions. Two-way tables generally give row and column headers, as standard terminology implies, so why not label columns on the top?

We are going to look at examples of the second and third kinds after some brief comments on Stata syntax for changing what goes on what axis.

4   Stata syntax

Given an interest in experimenting, how is this to be done? With `twoway` graphs, the easiest way to start is with `yscale(alt)` or `xscale(alt)`, which flips axes so that the other nondefault axis is used for title, labels, and ticks. With a Stata standard, consider

```
. sysuse auto
(1978 Automobile Data)
. scatter mpg weight, yscale(alt)
(graph omitted)
. scatter mpg weight, xscale(alt)
```

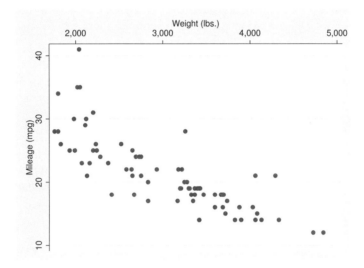

Figure 1: A standard scatterplot, except that x-axis labels are shown at the top

The second `scatter` command is more interesting and the result is shown as figure 1. It may strike most people as odd, but is this more than a reaction to a conventional choice being set aside? Let me spell it out: the origin $(0, 0)$ is not explicit on this graph; that is, neither zero `mpg` nor zero `weight` are near observed values, and no line on the graph is a true axis for which y or x is 0. To be sure, in this and many other graphs, there is a case that forcing Stata to show the origin might help many readers to think about the data, and that can be done by spelling out `yscale(r(0 .))` and `xscale(r(0 .))`.

The `alt` suboption here evidently stands for `alternate` and suffices for simple flipping of axes. The more elaborate machinery documented at [G-3] **axis_choice_options** is prominently documented, but not always needed. For example, using the `subtitle()` option is often the most direct method to put a title on the top. The subtitle can be moved around the graph by using its suboptions. It is easy to overlook the extra title options `l1title()`, `l2title()`, `r1title()`, `r2title()`, `t1title()`, `t2title()`, `b1title()`, and `b2title()`. Here `l`, `r`, `t`, and `b` stand for left, right, top, and bottom.

These title options come in pairs: 1 and 2 indicate outer and inner titles. Long-time Stata users may recollect the very similar options in the old graphics of Stata 7 and earlier.

5   Multiple quantile plots

Quantile plots have long been a favorite of mine. They plot the ordered values (order statistics or quantiles) from a distribution of one variable against the so-called plotting positions, in effect, estimates of the associated cumulative probability. For discussions and references, see Cox (1999, 2005). Stata implementations include the official `quantile` command (see [R] **diagnostic plots**) and my more general `qplot` command (`quantil2` was an earlier name of the latter). For the most recent update of `qplot`, download from the files for Cox (2012a).

`qplot` can be used to show the distributions of two or more variables, but that makes most sense when variables are measured in the same units and are broadly comparable in magnitude. However, a helpful, broad exploratory view of a dataset may be obtained by looking at several quantile plots together. In essence, that just implies cycling over a `qplot` call for each of several variables, letting each `qplot` find its own y-axis scale.

There are many ways to do this. The `multqplot` command distributed with this column is certainly indicative rather than definitive. It loops over a numeric *varlist* and uses `graph combine` to show the resulting graphs in one image. A formal statement of its syntax is given in a later section.

Let's see an example first. We return to the `auto.dta` example. `make` is a string variable, but the other 11 variables are numeric, and so typing

```
. multqplot price-foreign
```

shows all the univariate distributions (figure 2).

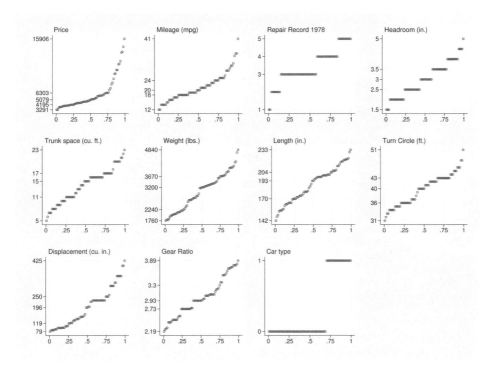

Figure 2: Multiple quantile plot of the numeric variables in `auto.dta`; the quantiles of each variable are plotted against the equivalent fractions of the data, calculated as (rank − 0.5)/sample size

Coincidentally, this example with 11 graphs is about the limit of what is comfortably seen on a modestly sized monitor. It is also true that even when reduced to about 1/12th of the usual graph size, a quantile plot still shows much interesting and intelligible detail. Those faced with hundreds or thousands of variables would need a different strategy. For example, arranging to see numerous quantile plots in a slide show is likely to be more practical.

Several choices in the design of the individual quantile plots shown by `multqplot` reflect a desire to make the most of a small space:

1. The variable information (a variable label or, in its absence, a variable name) goes on the top. Left-to-right orientation makes it more readable.

2. That frees up the y axis to some extent, and the default is to show the values of five summary values: minimum, lower quartile, median, upper quartile, and maximum. As grid lines are also shown by default for probabilities 0, 0.25, 0.5, 0.75, and 1, the combined effect is to hybridize quantile and box plots (compare especially Parzen [1979]). The box corresponding to 0.25, 0.5, and 0.75 quantiles is not, however, emphasized visually; but it would be possible to program having

two sets of grid lines with different thicknesses. Conversely, this plot shows much more information than a conventional box plot in so far as all quantiles are shown. If the same value is also reported as two or more summary quantiles, then fewer than five values are shown on the y axis. This will always happen with binary variables—for example, `foreign` in this dataset—and may easily happen with other variables with relatively few distinct values.

3. Although the default display of quantiles has just been discussed in detail, any set of quantiles may be selected to show numerically on the y axis. My advice is not to try to show too many, because they will be difficult to read.

4. The default x-axis title, "fraction of the data", would be the same for all individual `qplot`s and is suppressed partly to avoid repetition and partly because of lack of space. Putting an explanation in the text caption for the graph, as here, is thus natural.

`multqplot` makes some discreet use of grid lines. Fashion has moved away from grid lines recently in statistical graphics, but they can be useful; see Cox (2009b) for more discussion.

Further advocacy of quantile plots is not the main purpose of this column, and my previous papers in that vein are easily accessible. But those to whom quantile plots are unfamiliar might note some of their advantages:

1. They show well any outliers, gaps, or granularity, which is useful in data checking or assessment of data quality.

2. They scale well over a large range of possible sample sizes.

3. They entail a minimum of arbitrary choices compared with, say, bin origin and bin width for histograms or kernel type and width for kernel density estimation.

4. They signal behavior that might be awkward in later modeling of the data, whether variables are responses or predictors.

5. They behave reasonably with ordinal or binary variables. (They cannot do anything other than reflect the numerical coding of nominal variables.)

That said, showing small multiples for any univariate distribution plot is easy to copy for other displays, and `multqplot` could serve as a starting point for programmers. When you code alternatives, it may prove useful to rethink the style of the standard display.

6   Table-like graphs

Another favorite theme of mine in these columns has been the idea that the conventional distinction between tables and graphs may usefully be subverted. For much more discussion and some literature references, see Cox (2008). In Stata, it is much easier to make graphs more table-like than to make tables more graph-like. Although it is clearly a cosmetic detail, moving x-axis information to the top can make graphs even more table-like by mimicking the convention of column headers.

To see this in practice, we will revisit an example used in the documentation; see [R] **kappa**. The data concern the structure of agreement and disagreement between two radiologists in the diagnosis of a sample of tissues.

```
. webuse rate2
(Altman p. 403)

. kap rada radb, tab
```

Radiologis t A's assessment	Radiologist B's assessment				Total
	Normal	benign	suspect	cancer	
Normal	21	12	0	0	33
benign	4	17	1	0	22
suspect	3	9	15	2	29
cancer	0	0	0	1	1
Total	28	38	16	3	85

Agreement	Expected Agreement	Kappa	Std. Err.	Z	Prob>Z
63.53%	30.82%	0.4728	0.0694	6.81	0.0000

We can make the table graphical without losing any vital information. `tabplot` (Cox 2004) uses the simple idea of a two-way bar chart to represent cell frequencies from a table like that above. The most up-to-date version of `tabplot` may be installed on your system by typing `ssc install tabplot`; that version is needed to replicate the results here. The command

```
. tabplot rada radb, showval
```

gives a close replica of the core of the table (figure 3).

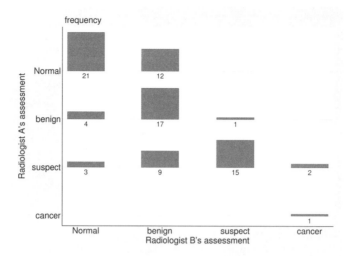

Figure 3: Two-way bar chart showing agreement and disagreement in diagnosis of two radiologists: bars are proportional in height to frequencies, and frequencies (cell counts) are shown below each bar

We can go further. First, there is enough space to show more detail. Let's show not only the frequencies but also those frequencies as a percentage of the total. In this case, the calculation is simple, although in many other datasets, you would need to watch for missing values.

```
. by rada radb, sort: generate freq = _N
. generate percent = 100 * freq/_N
```

`tabplot`'s `showval` option has a related option with the same name that takes arguments, including the name of a variable whose values are to be shown. (Rather, that is how it appears to the user: the simple programming of this subterfuge is explained in Cox [2009a].) The variable constructed for that purpose is a concatenation of values, spaces, and a percent sign. A `format()` suboption is used to round the percentages. (Showing percentages is itself another convention that usually works well, but there are alternatives. Showing numbers expressed per thousand [per mille] can even save a character [12.3 becomes 123], at the small cost of a little explanation.)

```
. generate show = string(freq) + "    " + string(percent, "%2.1f") + "%"
```

The final flourish is to flip the x-axis information to the top. At the same time, the default subtitle—in this case, `Frequency`—is just suppressed. Other choices are possible, but in this case, it seems simplest to transfer such explanation to the text caption as given below the graph. The result is figure 4.

```
. tabplot rada radb, showval(show) xscale(alt) subtitle("")
```

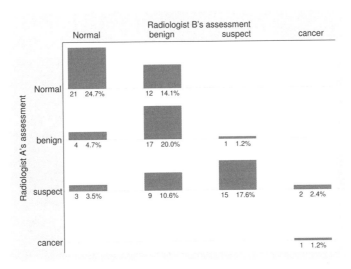

Figure 4: Two-way bar chart showing agreement and disagreement in diagnosis of two radiologists: Bars are proportional in height to frequencies, and frequencies are shown as both cell counts and percentage of the table total. *x*-axis labels have been moved to the top.

This is a relatively simple table. A conservative view that the graph does not add much to what is evident from the table is defensible, but the graph might prove as interesting or more so to some readers. Often tables are much more complicated, however, and then a graph may really help a reader to see structure more easily. But to return to the main idea of this column: Is the second graph preferable to the first?

7 Syntax of multqplot

multqplot *varlist* [*if*] [*in*] [, <u>all</u>obs <u>q</u>uantile(*numlist*) varnames
    qplot_options combine(*combine_options*)]

7.1 Description

multqplot combines and shows quantile plots for each of the numeric variables in *varlist* by using qplot, which must also be installed.

By default,

- plots are shown only for those observations for which every variable in *varlist* is nonmissing;

- each variable is explained by a `subtitle(, place(w))` at the top of each quantile plot, which shows the variable label or the variable name if no variable label is defined;

- quantiles for 0(25)100% of the data are shown by horizontal labels on the y axis;

- fractions of the data corresponding to those quantiles are shown by horizontal labels on the x axis;

- grid lines are shown matching y- and x-axis labels;

- titles are blank on both y and x axes.

Variations from the basic design may be obtained by particular option choices.

7.2   Options

`allobs` specifies that all observations specified are to be used as far as possible, regardless of the structure of missing values. The default is to use only those observations with nonmissing values on all the variables specified.

`quantile(`*numlist*`)` specifies a numlist of percentages between 0 and 100 (inclusive) that select particular quantiles to be calculated and shown as labels on the y axis. The numbers 0 and 100 are interpreted as the sample minimum and maximum, respectively. Quantiles are calculated with `summarize` (see [R] **summarize**), if possible, and with `_pctile` (see [D] **pctile**) otherwise.

`varnames` specifies that variable names should always be used to indicate which variable is being plotted. This option is provided for the circumstance in which at least one variable label is so long that labels should not be shown.

qplot_options are options of `qplot` used to tune individual quantile plots; see `help qplot`.

`combine(`*combine_options*`)` specifies any of the options documented in [G-2] **graph combine** that are used to tune the combination or joint display of the quantile plots. Note that this includes options for saving the graph to a file (see [G-3] ***saving_option***).

7.3   Remarks

As usual, this program is at best indicative, not definitive. Some might prefer that variables that are categorical by some definition or declaration be shown in some other way, say, by a bar chart. That is deliberately not an option here, but readers are welcome to clone the code and write their own variant program.

8 Conclusion

Appealing and effective graphics arises from a small series of good large decisions and a large series of good small decisions, most of which are to accept program defaults. This column has focused on one of the small decisions, placement of axis information. You are encouraged to be mildly subversive when it helps! Minute tweaks, such as putting y-axis information on the right or on the top or putting x-axis information on the top, often are useful. Indeed, they may already be someone else's orthodoxy.

9 References

Bochner, S. 1966. *The Role of Mathematics in the Rise of Science*. Princeton, NJ: Princeton University Press.

Boyer, C. B. 1956. *History of Analytic Geometry*. New York: Yeshiva University.

Brinton, W. C. 1914. *Graphic Methods for Presenting Facts*. New York: Engineering Magazine Company.

Cajori, F. 1929. *A History of Mathematical Notations. Volume II: Notation Mainly in Higher Mathematics*. Chicago: Open Court.

Court, A. 1987. Ten proposed rules of numerical diagrams. *Eos, Transactions, American Geophysical Union* 68: 1642.

Cox, N. J. 1999. gr42: Quantile plots, generalized. *Stata Technical Bulletin* 51: 16–18. Reprinted in *Stata Technical Bulletin Reprints*, vol. 9, pp. 113–116. College Station, TX: Stata Press.

————. 2004. Speaking Stata: Graphing categorical and compositional data. *Stata Journal* 4: 190–215.

————. 2005. Speaking Stata: The protean quantile plot. *Stata Journal* 5: 442–460.

————. 2008. Speaking Stata: Between tables and graphs. *Stata Journal* 8: 269–289.

————. 2009a. Stata tip 79: Optional arguments to options. *Stata Journal* 9: 504.

————. 2009b. Stata tip 82: Grounds for grids on graphs. *Stata Journal* 9: 648–651.

————. 2012a. Software update: gr42_6: Quantile plots, generalized. *Stata Journal* 12: 167.

————. 2012b. Speaking Stata: Transforming the time axis. *Stata Journal* 12: 332–341.

Klass, G. M. 2012. *Just Plain Data Analysis: Finding, Presenting, and Interpreting Social Science Data.* 2nd ed. Lanham, MD: Rowman & Littlefield.

Kline, M. 1972. *Mathematical Thought from Ancient to Modern Times.* New York: Oxford University Press.

Parzen, E. 1979. Nonparametric statistical data modeling. *Journal of the American Statistical Association* 74: 105–131.

Playfair, W. H. 1801a. *Commercial and Political Atlas: Representing, by Means of Stained Copper-Plate Charts, the Progress of the Commerce, Revenues, Expenditure, and Debts of England during the Whole of the Eighteenth Century.* 3rd ed. London: Wallis.

⸻. 1801b. *The Statistical Breviary: Shewing, on a Principle Entirely New, the Resources of Every State and Kingdom in Europe; Illustrated with Stained Copper-Plate Charts, Representing the Physical Powers of Each Distinct Nation with Ease and Perspicuity: to Which is Added, a Similar Exhibition of the Ruling Powers of Hindoostan.* London: Wallis.

⸻. 2005. *The Commercial and Political Atlas and Statistical Breviary.* Cambridge University Press: Cambridge. Edited by H. Wainer and I. Spence.

Stillwell, J. 2010. *Mathematics and Its History.* 3rd ed. New York: Springer.

About the author

Nicholas Cox is a statistically minded geographer at Durham University. He contributes talks, postings, FAQs, and programs to the Stata user community. He has also coauthored 15 commands in official Stata. He wrote several inserts in the *Stata Technical Bulletin* and is an editor of the *Stata Journal*.

The Stata Journal (2013)
13, Number 3, pp. 640–666

Speaking Stata: Trimming to taste

Nicholas J. Cox
Department of Geography
Durham University
Durham, UK
n.j.cox@durham.ac.uk

Abstract. Trimmed means are means calculated after setting aside zero or more values in each tail of a sample distribution. Here we focus on trimming equal numbers in each tail. Such trimmed means define a family or function with mean and median as extreme members and are attractive as simple and easily understood summaries of the general level (location, central tendency) of a variable. This article provides a tutorial review of trimmed means, emphasizing the scope for trimming to varying degrees in describing and exploring data. Detailed remarks are included on the idea's history, plotting of results, and confidence interval procedures. Examples are given using astronomical and medical data. The new Stata commands trimmean and trimplot are also included.

Keywords: st0313, trimmean, trimplot, trimming, means, medians, midmeans, Winsorizing, robust, resistant, graphics

1   Introduction

Trimmed means are means calculated after setting aside some values in one or both tails of a sample distribution. In the simplest and most common case, the same percent or number is set aside in each tail, omitting equal numbers of lowest and highest values.

The idea of trimming has been reinvented repeatedly since the 18th century. Trimmed means have been prominent as one of the simpler methods within the field of robust statistics for over 50 years, since their reintroduction by J. W. Tukey (1960), W. J. Dixon (1960), and others. The idea of trimming binds means and medians together in a wider family: the mean, strictly speaking, is the mean with no values trimmed, while the median is the mean with all values trimmed except the one or two values that define the median. Intermediate degrees of trimming offer varying compromises between the urge to use all the information in the data and any need to discount extreme values that may appear unreliable.

Trimming before averaging is easy to understand and to explain to general scientific audiences. Trimmed means are likely to be useful as a cautious check on means or as an alternative summary when using means seems dubious or even dangerous. Why then are they not more widely used? Lack of detailed explanations and implementations may be one answer, and this article addresses that lack for Stata users. Two earlier user-written programs will be discussed later, but in general, Stata has lagged behind other statistical software in this field: "a trimmed mean" was added to BMDP in 1977 (Hill and Dixon 1982, 378).

Trimmed means may be based on trimming differently in each tail, including the case of trimming in one tail only. Staudte and Sheather (1990), for example, first introduce trimmed means in terms of trimming only in the right tail when estimating the scale of an exponential distribution. Here we focus on trimming symmetrically.

The new commands `trimmean` and `trimplot` are included in this article. Discussion will follow a detailed explanation of the basic statistics, together with historical remarks.

Some comments follow first on the literature. The reviews by Dixon and Yuen (1974) and Rosenberger and Gasko (1983) remain clear and helpful on both specific details and wider context. Both have often been overlooked in later surveys for no obvious good reason. Kafadar (2003) gives an excellent concise review of Tukey's work on robustness. Stigler (2010) gives a light and brisk historical perspective on robust statistics in general.

Introductory or intermediate texts featuring trimmed means include Breiman (1973, 244), Mosteller and Tukey (1977, 34–36), Dixon and Massey (1983, 380–382), Siegel (1988, 66–68), Helsel and Hirsch (1992, 7), Venables and Ripley (2002, 122), van Belle et al. (2004, 276–277), Rice (2007, 397), Sprent and Smeeton (2007, 461–465), Reimann et al. (2008, 43), and Feigelson and Babu (2012, 110).

2 Definitions

2.1 Simplest case

The order statistics of a sample of n values of a variable y are defined by

$$y_{(1)} \leq y_{(2)} \leq \cdots \leq y_{(n-1)} \leq y_{(n)}$$

so that $y_{(1)}$ is the smallest value and $y_{(n)}$ is the largest.

The method for trimmed means at its simplest is to set aside some fraction of the lowest-order statistics and the same fraction of the highest-order statistics and then to calculate the mean of what remains, thus providing some protection against possible stretched tails or outliers in a sample. For example, suppose $n = 100$, and we set aside 5% in each tail, namely, $y_{(1)}, \ldots, y_{(5)}$ and $y_{(96)}, \ldots, y_{(100)}$. We can then take the mean of $y_{(6)}, \ldots, y_{(95)}$.

For such a definition, see Tukey and McLaughlin (1963, 336), Bickel (1965, 848), Huber (1981, 57–58), Lehmann (1983, 360), Rosenberger and Gasko (1983, 307–308), Hampel et al. (1986, 178), Staudte and Sheather (1990, 104), Barnett and Lewis (1994, 79), Miller (1986, 29), David and Nagaraja (2003, 213), Jurečková and Picek (2006, 67), Pearson (2011, 228, 267), or Wilcox (2003, 62–63; 2009, 26; 2012a, 55; 2012b, 25).

The 0% trimmed mean is thus just the usual mean.

By courtesy, or as a limiting case, the 50% trimmed mean is taken to be the median. The small detail here is that trimming exactly half the values in each tail will leave no values at all; hence, the courtesy exercised in leaving the one or two values required to determine the median. (Averaging the two central values whenever n is even to calculate the median is explained to mathematical audiences as a convention and to nonmathematical audiences as a rule.)

One other trimmed mean has often been given a special name. The 25% trimmed mean has been called the "midmean", that is, the mean of the middle half of the data (Tukey [1970a; 1970b, 168], adopting earlier scientific usage, on which see Tukey [1986, 871]); the "interquartile mean" (for example, Tilanus and Rey [1964]; Erickson and Nosanchuk [1977, 40; 1992, 44]); and the "quartile-discard average" (Daniell 1920).

A more general rule is that the lowest value included in the calculation of the $p\%$ trimmed mean is $y_{(g+1)}$, where $g = \lfloor n\,p/100 \rfloor$, and the highest value included is thus $y_{(n-g)}$.

The very useful floor notation, $\lfloor\;\rfloor$, here specifies rounding down to the nearest integer. Incidentally, almost all the literature on trimmed means uses [] with the same meaning. Despite a lengthy pedigree, that notation needs to be explained repeatedly: many readers might disregard it as merely a standard use of brackets. See Cox (2003) for more discussion and further references on floors and ceilings.

2.2   Weighting

Some authors use a more elaborate definition of trimmed means in which some values may be given fractional weights. See Andrews et al. (1972, 7, 31), Stigler (1977, 1060), Kleiner and Graedel (1980, 706), Huber (1981, 57–58), Rosenberger and Gasko (1983, 311), Barnett and Lewis (1994, 79), Huber and Ronchetti (2009, 57–58), or Wilcox (2012a, 55).

The precise rule is usually that $\lfloor n\,p/100 \rfloor$ values are removed in each tail, and the smallest and largest remaining values are assigned weight $1 + \lfloor n\,p/100 \rfloor - n\,p/100$. So, for example, given $n = 74$ and percent $5/100$, their product is 3.7. Rounding down gives 3 and so we work with $y_{(4)}, \ldots, y_{(71)}$. However, $y_{(4)}$ and $y_{(71)}$ are assigned weight $4 - 3.7 = 0.3$, and $y_{(5)}, \ldots, y_{(70)}$ assigned weight 1. Then a weighted mean is taken.

The idea underlying this alternative definition appears twofold: $p\%$ should mean precisely that, and the result of trimming should vary as smoothly as possible with p. Rosenberger and Gasko (1983, 310–311) explain this especially clearly with two helpful diagrams.

The difference is partly a matter of taste. But always using weights that are 1 or 0 is appealingly simple and appears entirely adequate for descriptive and exploratory uses. Moreover, any fine structure that results from the inclusion and exclusion of particular values as trimming proportion varies is likely to be trivial or part of what we are watching for. Either way, there is little loss.

2.3   Number instead of percent trimmed

In some situations, it is more natural to specify trimming in terms of the number of values trimmed rather than the percent. For example, trimming or truncating procedures have been used in combining the scores of a panel of judges in various sports to discourage or discount bias for or against competitors. Here the rules might require, for example, trimming the highest and lowest values.

Focusing on the number trimmed allows a slightly different definition of trimmed means. We can describe the order statistics $y_{(1)} \leq y_{(2)} \leq \cdots \leq y_{(n-1)} \leq y_{(n)}$ using the idea of depth (for example, Tukey [1977]). Depth is defined as 1 for $y_{(1)}$ and $y_{(n)}$, 2 for $y_{(2)}$ and $y_{(n-1)}$, and so forth: it is the smaller number reached by counting inward from either extreme $y_{(1)}$ or extreme $y_{(n)}$ toward any specified value. So the depth of $y_{(i)}$ is the smaller of i and $n - i + 1$.

A trimmed mean may be defined for any particular depth as the mean of all values with that depth or greater. Thus the trimmed mean for depth 1 is the mean of all values. The trimmed mean for depth 2 is the mean of all values except those of depth 1, that is, all values except for the extremes. The trimmed mean for depth 3 is the mean of all values except those of depths 1 and 2, and so forth.

The highest depth observed for a distribution occurs once if n is odd and twice if n is even; either way it labels values whose mean is the median. Thus, again, trimmed means range from the mean to the median.

2.4   Symmetry or asymmetry?

Whatever the precise definition, trimming the same number of order statistics in each tail is arguably based on a symmetry assumption—if not that the distribution of interest is approximately symmetric, then that the chances of contamination are approximately equal in either tail. Certainly, estimation of location (level, central tendency) is easy to think about whenever the underlying distribution is symmetric (and easier still if it is unimodal). Then estimators of location can typically be thought of as aimed at precisely the same target, the middle or center of the distribution.

The opposite argument is that the estimand is whatever the estimator points to. As Tukey (1962, 60) urged, "We must give even more attention to starting with an estimator and discovering what is a reasonable estimand, to discovering what is it reasonable to think of the estimator as estimating". A similar point of view has been elaborated formally in considerable detail by Bickel and Lehmann (1975) and informally with considerable lucidity by Mosteller and Tukey (1977, 32–34). There is also an elementary version. Using, say, sample median or geometric mean to estimate the corresponding population parameter makes sense regardless of whether the underlying distribution is symmetric, and the same goodwill can be extended to sample trimmed means, which are regarded as estimators of their population counterparts.

All that said, symmetric trimmed means are unlikely to be ideal for strongly asymmetric distributions. So-called J-shaped distributions such as the exponential or Pareto are examples. As remarked in the *Introduction*, there has been work on asymmetric trimmed means, but that is not discussed further in this article.

2.5 Confidence intervals

Trimmed means can be given confidence intervals. The approach here follows Tukey and McLaughlin (1963). Note also Dixon and Tukey (1968) as the sequel to that article. For a one-sentence summary, see Huber (1972, 1053–1054). For lucid textbook accounts, see Staudte and Sheather (1990, 98), Miller (1986, 30–31), Huber and Ronchetti (2009, 147–148), or Wilcox (2003, 126–132; 2009, 98–99, 127–128, 150–151; 2010, 153–154; 2012a, 57–61, 111–114; 2012b, 153–159).

Suppose we have n values and trim g in each tail and we seek *level*% confidence intervals (for example, *level* = 95). We need first a Winsorized standard deviation. Winsorizing is replacing values in each tail by the next inward value; that is, $y_{(1)}, \ldots, y_{(g)}$ are each replaced by $y_{(g+1)}$, and $y_{(n-g+1)}, \ldots, y_{(n)}$ are each replaced by $y_{(n-g)}$ before calculation, so long as $g \geq 1$. Let sd_W denote the standard deviation of the Winsorized values. Then intervals are mean $\pm$ (t multiplier $\times$ sd_W) / $\{\sqrt{n}\,(1 - 2g/n)\}$, where the t multiplier in Stata terms is `invttail($n$ - 2*g - 1, (100 - *level*)/200)`. If the latter expression looks too much like a strange incantation, consider an example such as $n = 100$, $g = 5$, and *level* = 95:

```
. display invttail(100 - 2*5 - 1, (100 - 95)/200)
1.9869787
```

Note that the `trimmean` command uses `summarize` to calculate the standard deviation; thus, as documented in [R] **summarize**, the divisor before rooting is $n - 1$. The fraction of values used in the trimmed mean $1 - 2g/n$ is calculated from the number actually used, not from any percent trimming specified.

This approach does not in the limit as trimming approaches 50% give reasonable confidence intervals for the median, because the number of degrees of freedom in this method approaches 0. `trimmean` declines to cite confidence intervals for the median; otherwise, obtaining intervals for large trimming fractions is left to the judgment of the user.

As another approach to confidence intervals, bootstrapping is quite attractive. Efron and Tibshirani (1993) and Davison and Hinkley (1997) discuss bootstrapping trimmed means. Although all results are returned in a matrix, `trimmean` also saves each trimmed mean separately as a convenience. However, bootstrapping necessarily implies that wild values could be selected repeatedly in a bootstrap sample, so some individual trimmed means could be much less resistant than the mean based on the sample as a whole. The converse is also true.

2.6 Metric trimming

A different definition of trimmed means, often called metric trimming, yields means of values satisfying some constraint on the absolute deviation from the median $|y - \text{median}(y)| =: d$. The name "metric" echoes Bickel (1965), Kim (1992), and Venables and Ripley (2002, 122). None of those cited the earlier work of Short (1763), similar in spirit, except that he worked with $|y - \text{mean}(y)|$.

Two simple merits of this definition deserve mention. Like the usual definition, it defines a family spanning the median and mean as extremes and including intermediate compromises. Specifying allowed deviations on the scale of the variable may make much sense to working scientists accustomed to thinking about their measurements.

Thus $d = 0$ identifies data points equal to the median. (A small detail here is that quite possibly, $d = 0$ identifies no data points at all: that will necessarily happen whenever n is even and the median is calculated as the mean of two different values.)

At the other extreme, such a trimmed mean equals the mean so long as d exceeds the largest possible absolute deviation, the larger of $\text{median}(y) - y_{(1)}$ and $y_{(n)} - \text{median}(y)$.

Lest readers confuse this with other procedures, this is not an iterative calculation. The overall median is calculated just once; there is no cycling such that the median is redefined to be the median of those values $\leq d$.

Metric trimming can be combined with trimming based on order statistics (for example, Hampel [1997, 150] and Olive [2001]), but only the simplest flavor is supported by `trimmean`. See also Huber (1964) for a brief mention and Hampel (1985) for broader discussion.

3   Historical remarks

The idea of a trimmed mean is quite old. For some related history, see Stigler (1973, 1976), Harter (1974a,b), Hampel et al. (1986, 34–36), and Barnett and Lewis (1994, 27–31). The episodes identified here seem best thought of as independent inventions and not as evidence of a continuous thread of thought intermittently made visible.

Throughout several centuries, trimmed means have been one of several practices in science and, typically, only an occasional practice. Pooling all measurements and calculating a single mean has at best been one approach and only slowly came to be regarded as one standard. Choosing the best measurement from several as a matter of judgment was, and remains, an alternative often used both in science and in everyday life. Some scientists and a few statisticians have focused on "rejection of outliers", that is, identification either by judgment or by some formal rule of outlying values not to be trusted. This might be included under trimming in a broad sense. However, it seems best to distinguish clearly: Trimmed means are based on choosing a rule for trimming, whether a percent or number to be trimmed or a maximum allowed deviation. Rejection of outliers is based on looking at the data, deciding which, if any, values need to be rejected, and then averaging what remains.

Five case studies follow from the literature, with absolutely no claim to completeness. Biographical vignettes on the main individuals follow the References.

3.1   James Short and the 1761 transit of Venus

James Short (1763) used a form of what is now called metric trimming in 1763 for estimating the sun's parallax based on observations of the transit of Venus across the face of the sun, namely, taking the mean of values closer than some chosen distance from the mean of all. The parallax here is the angle subtended by the earth's radius, as if viewed and measured from the surface of the sun. The units are seconds of a degree. Note that repeating Short's (1763) calculations points up small errors in his arithmetic. For much more on measuring the transit of Venus in 1761 (and 1769) as a major research program in astronomy, see Woolf (1959) or Wulf (2012). Woolf (1959, 147) comments: "One of the factors that had rendered Short's results so homogeneous had been the rather judicious series of alterations which he had made in the original data concerning longitude and time of contact at various stations". Short's (1763) line, however, was that he was fixing the mistakes of others.

Stigler (1977) included Short's (1763) data in an evaluation of robust estimation methods with real data.

Short (1763) provides 53 measurements on page 310 of his article. These are datasets (1) to (3) in Stigler (1977, 1074). Short (1763) first averages all 53 and gets 8.61; then all 45 within 1 of that mean and gets 8.55; then all 37 within 0.5 of that mean and gets 8.57. Then he takes the mean of all 3 means and gets 8.58. In effect, his final mean is weighted according to deviations from the initial overall mean.

Similarly, Short (1763) provides 63 measurements on page 316 of his article. These are datasets (4) to (6) in Stigler (1977, 1074). Short (1763) first averages all 63 and gets 8.63; then all 49 within 1 of that mean and gets 8.50; then all 37 within 0.5 and gets 8.535. The mean of all 3 means is 8.55.

Short's (1763) data on page 325 of his article are datasets (7) and (8) in Stigler (1977, 1074). The mean of 21 values in the first set, for the Cape of Good Hope, is 8.56. All 29 values are within 0.2 of that. The mean of 21 values in the second set, for Rodrigues, is 8.57; the same mean is obtained for all 13 within 0.2.

3.2   A French custom

An anonymous writer (identified by Stigler [1976] as Joseph Diaz Gergonne, 1771–1859) included an example of trimmed means in a discussion of how to calculate means (Anonymous 1821, 189): "For example, there are certain provinces of France where, to determine the mean yield of a property of land, there is a custom to observe this yield during twenty consecutive years, to remove the strongest and the weakest yield and then to take one eighteenth of the sum of the others" (translation in Huber [1972, 1043]).

3.3   Mendeleev on metrology

Mendeleev (1895) (reference in Harter [1974b, 241]) reported his method "to evaluate the harmony of a series of observations that must give identical numbers, namely, I divide all the numbers into three, if possible equal, groups (if the number of observations is not divisible by three, the greatest number is left in the middle group): those of greatest magnitude, those of medium magnitude, and those of smallest magnitude; the mean of the middle group is considered the most probable ... and if the mean of the remaining groups is close to it ... the observations are considered harmonious". Thus Mendeleev used a 1/3, roughly 33%, trimmed mean.

3.4   Daniell's theoretical treatment

Daniell (1920) gave an elegant and pathbreaking general treatment of statistics that are linear combinations of the order statistics, including various estimators of location and scale. It was apparently inspired by a reading of Poincaré's *Calcul des probabilités* (1912). Daniell derived optimal weighting functions and gave the first mathematical treatment of the trimmed mean. However, his article had essentially no impact before its rediscovery by Stigler (1973). Its placement in a journal rarely read by statisticians cannot have helped.

3.5   Tukey and modern robust statistics

Tukey (1960) surveyed the problem of location estimation when data are likely to come from distributions heavier tailed than the normal (Gaussian) in an outstanding article that was one of the founding documents of modern robust statistics. He combined a literature review with a report on his own work on the subject since the mid-1940s, some published partly as technical memoranda. In particular, Tukey (1960) showed that truncated means (his term in this article) calculated after dropping the same percent of the lowest and highest values offered considerable protection in the face of such distributions. Dixon (1960) also deserves credit for work in this territory.

The term "trimmed mean" was introduced shortly afterward by Tukey (1962). Names in earlier use include "truncated mean" (Tukey [1960], as above) and "discard average" (Daniell 1920). Dixon (1960) discussed using means of a censored sample. Talking of truncation or censoring raises the need to distinguish carefully between truncation or censoring of the data before they arrive and such truncation or censoring used deliberately in data analysis—reason enough for using the term "trimming" instead.

After 1960, trimmed means became an established method within the field of robust statistics, a field repeatedly surveyed and unified by monographs and textbooks and now covering many other statistical problems, including robust regression.

4   How much to trim in practice?

A simple and natural question not raised so far is how much to trim in practice. There is a simple and natural answer: that depends on your dataset, your ideas about generating processes, and your attitude to risk. But let us back up and consider more generally how we might evaluate trimmed means and make choices.

One of the advantages of trimmed means is that their behavior is easy to think about. Trimming $p\%$ in each tail of a distribution offers protection against (up to) the same fraction of dubious values in each tail. A more formal treatment would be phrased in terms of the idea of "breakdown point" (Donoho and Huber 1983). Whether these dubious values are called outliers, or something else, is partly a matter of taste and judgment. There is also some taste and judgment in trading off the protection afforded against dubious values against the loss of information incurred by ignoring values that may be genuine.

As with any other method, trimmed means may be studied theoretically, including by simulation from distributions deemed credible as generating processes, or empirically, by studying how trimmed means behave with real data. The broad advantages and disadvantages of each method are clear.

The monograph by Andrews et al. (1972) remains the most impressive compendium of simulation results for trimmed means (and other robust estimators of location). It is striking to learn from Hampel (1997) that only a small part of the project was ever written up. One important omission was what happens with skewed or asymmetric distributions. But the mass of previous simulation results, both in that volume and elsewhere, is increasingly redundant. It is now easier to simulate afresh using whatever underlying distributions and sample sizes appear pertinent to any particular project than to comb through the literature searching for relevant results.

Similarly, studies with other datasets still raise the question of judging which other results are pertinent to the current project. Such studies (for example, Stigler [1977] and Hill and Dixon [1982]) are often more positive than reports of simulation studies, implying in particular that mild trimming (loosely, of the order of 5–10%) may be all that is required in many cases.

In either case, there are severe selection problems. How do you reasonably sample from the space of possible distributions, whether theoretical or empirical? What part of that space is relevant to your project? These are difficult questions.

One tactic that seems underplayed is to use trimming percents across a wide range and see what happens. An obvious aid here is to plot trimmed mean versus number or percent trimmed or allowed deviation. Examples of such plots can be found in Rosenberger and Gasko (1983, 315) and Davison and Hinkley (1997, 122).

Then the possibilities include, but are not limited to, some leading cases:

1. Results are stable, whatever the trimming proportion. We can relax and just use means any way. This case is like a health check or machine service that found no problems.

2. Results are stable provided that you trim at least a certain proportion.

3. Results vary systematically with trimming proportion, from mean to median. Note that this is expected with most asymmetric or skewed distributions, regardless of whether outliers or heavy tails are present. (It is often forgotten that there are skewed distributions for which mean and median are identical or very close. For example, this is true of some binomial distributions and usually a good approximation for Poisson distributions.) In turn, there will be choices, including living with the fact; realizing that multiple descriptors—say, mean and median and perhaps others as well—may be advisable in reporting data; and considering an appropriate transformation or link function (in the jargon of generalized linear models, see [R] **glm**).

4. Something else that needs consideration. Trimmed means may here indicate a problem, but they do not promise to provide a solution.

`trimmean` is designed to make it easy to produce several trimmed means at once by specifying differing trimming percents, numbers, or allowed deviations from the median. The accompanying program `trimplot` provides graphical display of results. In fact, you might prefer to look at a graph from `trimplot` first. (For the syntax of both commands, see sections 7 and 8.)

5   Applications to real data

5.1   Short (1763) revisited

For a first example, we will revisit Short's (1763) measurements of parallax. His data are provided with the media for this issue in `short.dta`. The variables are `parallax`, measured in decimal seconds of a degree, and `page`, meaning page in the original article. (As already mentioned in section 3.1, there were two datasets on page 325.)

For a first look at data, I often turn to quantile plots as capable of showing both broad features and any unusual details and specifically to the program `qplot`. See Cox (1999, 2005) for general discussion and many references and the files associated with Cox (2012a) for code download. (A `search` for `qplot` in an up-to-date Stata may reveal a later version, depending on when you read this.)

```
. use short
. qplot parallax, by(page)
```

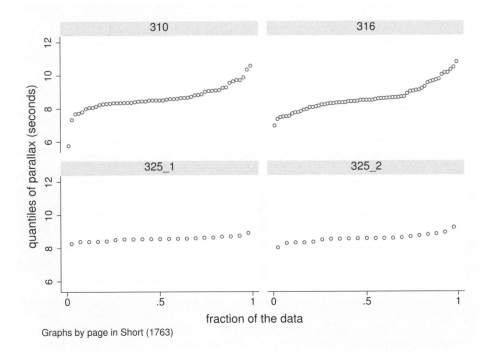

Figure 1: Quantile plots for subsets of parallax measurements in Short (1763)

Given four subsets, there is a choice between showing them separately (juxtaposed) with the `by()` option of `qplot` and showing them together (superposed) with its `over()` option. Both can be useful. We see immediately from figure 1 that the two smaller subsets are distinctly less variable than the two larger ones. No wild outliers are apparent in each case.

`trimplot` shows all possible trimmed means. Specifying `percent` as an option is useful for comparing subsets of differing sizes. Instead of using the `by()` option, we can use the `over` option to show results in a single panel. As with `qplot`, both choices are allowed. There is enough space to put the legend inside the plot region in the top right-hand corner. See figure 2.

```
. trimplot parallax, over(page) percent legend(position(1) ring(0) cols(1))
```

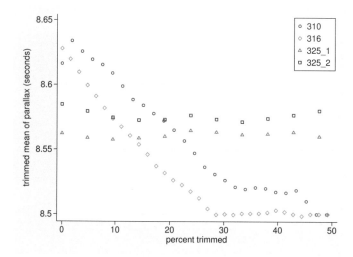

Figure 2: All possible trimmed means for subsets of parallax measurements in Short (1763)

The results show a simple contrast. Trimmed means are stable for the two smaller subsets but drift fairly systematically with trimming proportion for the two larger subsets. The graph quantifies agreement (all results shown are between 8.50″ and 8.63″ to 2 decimal places) and also disagreement (there appears to be some systematic difference between the two pairs). Were this a live issue, the results could now be taken back to the astronomical community for reflection and further analysis.

Graphs such as figure 2 show percent trimmed on the x axis. This encourages thinking in terms of what happens as we trim more and more. The opposite interpretation is also possible: what happens as we use more and more of the values in the data? That interpretation would be made easier by reversing the axis, which is just a standard `twoway` option call, `xsc(reverse)`.

To see the numbers, we can fire up `trimmean`. If desired, `trimmean` will show all possible trimmed means through its `number()` option, but a display of values for percents `0(5)50` appears to be adequate detail for many problems. The tabulation is suppressed here to save space.

```
. by page, sort: trimmean parallax, percent(0(5)50) format(%3.2f)
  (output omitted)
```

A final flourish with Short's (1763) data is to get closer to what Short (1763) actually did and trim metrically by using an idea of maximum allowed deviation. As mentioned already, Short (1763) worked with deviation from the overall mean, whereas we find the idea of working with deviation from the overall median much more appealing. Crucially, means can be pulled way off by any dubious values, and focusing only on values close to the mean may not be enough protection.

Here again a question arises about axis direction. To be consistent with the previous plot, in which trimming amount increases left to right, we reverse this plot's x axis scale. Partly to underline what is possible and partly to use the available space more fully, we use separate scales for each subset. See figure 3.

```
. trimplot para, by(page, xrescale) metric xsc(reverse)
```

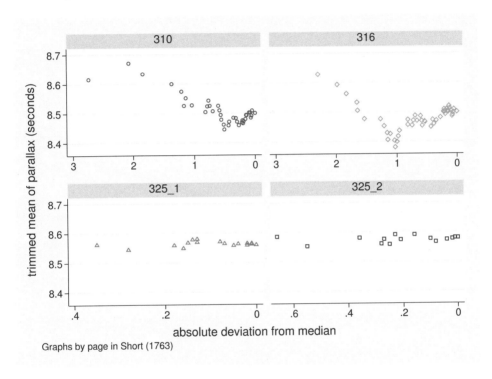

Figure 3: All possible metrically trimmed means for subsets of parallax measurements in Short (1763)

5.2   Chapman data

For a second example, we turn to a dataset on 200 men from Dixon and Massey (1983, 17–20) called the Chapman heart study data. (Details about Chapman were not recorded there.) This appears to be of fairly high quality. We pick four variables with quite different characters and units of measurement. For a basic view, a multiple quantile plot is obtained with `multqplot` (Cox 2012b). See figure 4.

```
. use chapman
. multqplot diastolic cholesterol height weight
```

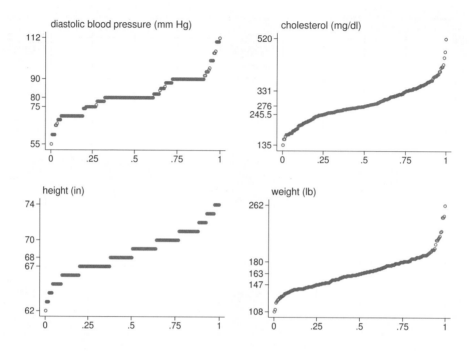

Figure 4: Quantile plots for diastolic blood pressure, cholesterol, height, and weight for 200 men in the Chapman heart study data

By default in this graph, 5 quantiles are labeled on the y axis: minimum, lower quartile, median, upper quartile, and maximum, or the 0, 25, 50, 75, and 100% points in a distribution. These are precisely the quantiles explicit in the most common kind of box plots. The difference from box plots is that quantile plots show all distinct values. They will necessarily, but unproblematically, blur into each other to some extent in many large datasets. Conversely, it is almost a definition that outliers will remain distinct.

Three broad features stand out from the multiple display, not all of which would be strongly evident from corresponding box plots or histograms. First, all variables are approximately symmetric to mildly right skewed. Second, there are no marked outliers. Third, diastolic blood pressure and height stand out as granular in detail with several ties at rounded values. The motive of the measures of avoiding spurious precision may easily be guessed, but the granularity is still strong. Diastolic blood pressure is often reported as 70, 80, or 90 mm Hg. Height is measured to the nearest inch. (Note: 1 inch = 25.4 mm.)

Given the diversity of values and units of measurements, separate `trimplots` are advisable. We loop over the four variables and then use `graph combine`. The result is shown in figure 5.

```
. foreach v in diastolic cholesterol height weight {
  2.          trimplot `v´, percent name(`v´)
> xlabel(0(5)50) xline(0 25 50, lcolor(gs10) lwidth(medium))
  3. }
. graph combine diastolic cholesterol height weight
```

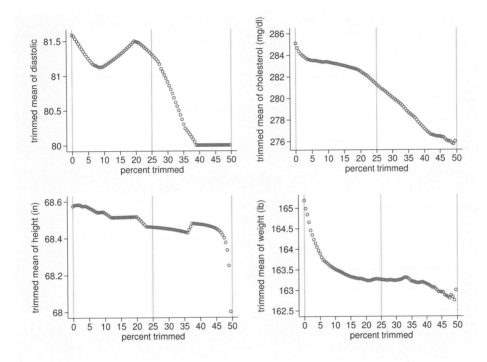

Figure 5: All possible trimmed means for diastolic blood pressure, cholesterol, height, and weight for 200 men in the Chapman heart study data

Grid lines have been added at 0, 25, and 50% trimming to flag the positions of the mean, midmean, and median. In detail, some linear segments of the traces for `diastolic` and `height` are evidently side effects or artifacts of the granularity of each distribution. Relatively smooth, approximately flat, or monotonic patterns are expectable when a distribution is symmetric or skewed and when there are many differing values.

For description or exploration, the main idea is that a plot of trimmed means could help underline that different summaries are close enough that one will be adequate, or that different summaries need to be quoted together, or indeed that the variable should be transformed. The midmean appears unjustly neglected as a simple descriptor. It has the merit that quartiles are widely familiar to users of statistics, so the idea can easily be explained. Having a memorable and evocative name also helps, even if it is pure superstition to suppose that a name is anything but a label.

For inference, how do confidence interval procedures work? As an experiment, I calculated 95% confidence intervals for the midmean using the Tukey–McLaughlin method and three standard `bootstrap` procedures. The script is too long to reproduce here but is available with the media for this issue as `trimming_ci.do`. I used 10,000 replications for bootstrapping. The random number seed is explicit in the do-file. Figure 6 shows good broad agreement, or for the skeptical or cynical, a reminder that different procedures usually give at least slightly different results.

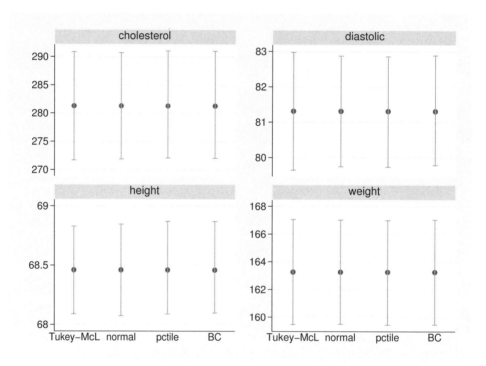

Figure 6: 95% confidence intervals for midmeans of diastolic blood pressure, cholesterol, height, and weight for the Chapman data, obtained by the Tukey–McLaughlin method and three bootstrap procedures

6   A note on other Stata implementations

The user-written program `iqr` by Hamilton (1991) calculates the 10% trimmed mean (only) as a sideline to other aims. His definition is the mean of values greater than the

10% percentile and less than the 90% percentile as calculated by [R] **summarize**, so results may often differ at least slightly from those calculated by `trimmean`.

The user-written program `robmean` by Ender (2009) calculates trimmed means according to the fraction trimmed (this is equivalent to the default of `trimmean` with the `percent()` option), together with some other quantities.

7 The trimmean command

7.1 Syntax

trimmean *varname* [*if*] [*in*],

    {<u>percent</u>(*numlist*) | <u>number</u>(*numlist*) | <u>metric</u>(*numlist*)} [<u>ceiling</u> <u>w</u>eighted ci

    <u>level</u>(#) <u>f</u>ormat(*format*) <u>g</u>enerate(*newvar*)]

    by ... : may also be used with `trimmean`; see help on `by`.

7.2 Description

`trimmean` calculates symmetric trimmed means as descriptive or inferential statistics for *varname*.

7.3 Options

percent(*numlist*) specifies percents of trimming for one or more trimmed means. Percents must be integers between 0 and 50 but otherwise can be specified as a *numlist*. Precisely one of `percent()`, `number()`, or `metric()` is required.

number(*numlist*) specifies numbers of values to be trimmed for one or more trimmed means. Numbers must be zero or positive integers less than half the number of observations available but otherwise can be specified as a *numlist*. Precisely one of `percent()`, `number()`, or `metric()` is required.

metric(*numlist*) specifies trimming such that means are of values within a specified absolute deviation of the median of a variable, say, y. Suppose `metric(0 100 200)` is specified. Then the means are means of values satisfying $|y - \mathrm{med}(y)| \leq 0, 100, 200$. Deviations must be zero or positive values but otherwise can be specified as a *numlist*. Precisely one of `percent()`, `number()`, or `metric()` is required.

ceiling specifies the use of `ceil()` rather than `floor()` in the calculation of ranks to be included. It is allowed with `number()` or `metric()` but ignored as irrelevant. This variation is occasionally suggested in the literature (for example, Huber in Andrews et al. [1972, 254]).

weighted implements a weighted variant explained in detail in the *Definitions* section in the help. It is allowed with number() or metric but ignored as irrelevant. This option may not be combined with ci.

ci specifies production of confidence intervals. This option may not be combined with weighted or metric(). For detailed discussion, see the *Definitions* section in the help.

level(*#*) specifies the confidence level, as a percentage, for confidence intervals. The default is level(95) or as set by set level; see [U] **20.7 Specifying the width of confidence intervals**.

format(*format*) specifies a numeric format for displaying trimmed means (and confidence limits when requested). The default is the display format of *varname*.

generate(*newvar*) specifies that an indicator (a.k.a. dummy) variable be generated with value 1 if an observation was included in the last trimmed mean calculated and 0 otherwise. The trimmed mean with the highest trimming percent or number or allowed deviation is always produced last, regardless of user input.

7.4   Stored results

trimmean stores the following in r():

Scalars
  r(tmean*#*)   each trimmed mean for percent or number *#* (for example, r(tmean5) for 5%) (with the metric() option, labeling is 1 upward, not with deviations specified)

Matrices
  r(results)   Stata matrix with columns as percents or numbers, number averaged, and trimmed means (and confidence limits when requested)

8   The trimplot command

8.1   Syntax

trimplot *varname* $\big[$ *if* $\big]$ $\big[$ *in* $\big]$ $\big[$, {over(*overvar*)|by(*byvar* $\big[$, *by_subopts* $\big]$)}
    percent metric mad *scatter_options* $\big]$

trimplot *varlist* $\big[$ *if* $\big]$ $\big[$ *in* $\big]$ $\big[$, percent metric mad *scatter_options* $\big]$

8.2   Description

trimplot produces plots of trimmed means versus depth or percent trimmed or deviation for one or more numeric variables. Such plots may help specifically in choosing or assessing measures of level and generally in assessing the symmetry or skewness of distributions. They can be used to compare distributions or to assess whether transformations are necessary or effective.

trimplot may be used to show trimmed means for one variable, in which case different groups may be distinguished by the over() or the by() option, or for several variables.

8.3   Options

over(*overvar*) specifies that calculations be carried out separately for each group defined by *overvar* but plotted in the same panel. over() is allowed only with a single variable to be plotted. over() and by() may not be combined.

by(*byvar* [, *by_subopts*]) specifies that calculations be carried out separately for each group defined by *byvar* and plotted in separate panels. Suboptions may be specified to tune the graphical display; see help on *by_option*. by() is allowed only with a single variable to be plotted. over() and by() may not be combined.

percent specifies that depth be scaled and plotted as percent trimmed, which will range from 0 to nearly 50 (a median cannot be based on no observed values, so 50 cannot be attained).

metric specifies that trimmed means be defined and plotted in terms of allowed absolute deviation from the median.

mad specifies metric trimming as above, but values will be plotted versus absolute deviation from the median / median absolute deviation from the median. The median (absolute) deviation (from the median) can be traced to Gauss (1816).

scatter_options are options of twoway scatter.

9   Conclusions

All careful users of statistics worry about how to handle awkward data. One well-documented kind of awkwardness consists of outliers and tails heavier than normal (Gaussian), a source of worry if techniques being used work best whenever some distribution is normal. Even if the modeling assumption is only that some conditional distribution be normal, marked nonnormality can still be awkward in various senses, not least in terms of how best to summarize and report such distributions.

Skepticism about the normal or Gaussian is often presented as a recent phenomenon, but it has deeper roots. Poincaré (1896, 149; 1912, 170–171) quoted a remark by Lippmann (Gabriel Lippmann 1845–1921, Nobel Prize for Physics 1908). This remark has often been misquoted or loosely translated, so here is a close translation from Mazliak (2012, 187): "[This distribution] cannot be obtained by rigorous deductions; many a proof one had wanted to give it is rough, among others the one based on the statement that the probability of the gaps is proportional to the gaps. Everyone believes it, however, as M. Lippmann told me one day, because the experimenters imagine it is a mathematical theorem, and the mathematicians that it is an experimental fact".

The field of robust statistics offers various solutions in this area. Trimmed means are one of the oldest and simplest methods for summarizing sample evidence on location as robustly as you wish. Why then are they not more frequently used? Speculation is easy. Perhaps they fall uncomfortably between teaching material and the research literature: they are too much of a complication or distraction to be included in many courses or texts and too simple or too well known to receive sustained focus in monographs on robust statistics. The larger problem for the latter is, naturally, how to model robustly the relationships between outcomes and predictors. The ever-elusive goal of a robust regression that is easy to understand, always reliable, and suitably fast seems likely to drive research for the indefinite future.

Whatever the precise diagnosis, I have focused here on providing constructive answers by way of a tutorial review and usable programs for trimmed means. The most distinctive emphases are including confidence interval procedures and emphasizing the scope for plotting results. Confidence intervals are often not mentioned in introductory accounts. The importance of plotting both raw data and results for trimmed means is also often understated. A more specific suggestion is that the midmean, the mean of the middle half of the data, seems unfairly neglected.

10　Acknowledgments

Rebecca Pope and Ariel Linden gave useful comments on earlier versions of `trimmean` and `trimplot`. David Hoaglin was very helpful in identifying citations in and around Tukey's work.

11　References

Aldrich, J. 2007. "But you have to remember P. J. Daniell of Sheffield". *Electronic Journal for History of Probability and Statistics* 3(2): 1–58.

Andrews, D. F., P. J. Bickel, F. R. Hampel, P. J. Huber, W. H. Rogers, and J. W. Tukey. 1972. *Robust Estimates of Location: Survey and Advances*. Princeton, NJ: Princeton University Press.

Anonymous. 1821. Probabilités. Dissertation sur la recherche du milieu le plus probable, entre les résultats de plusieurs observations ou expériences. *Annales de Mathématiques pures et appliquées* 12: 181–204.
http://www.numdam.org/item?id=AMPA_1821-1822__12__181_0.

Barnett, V., and T. Lewis. 1994. *Outliers in Statistical Data*. 3rd ed. Chichester, UK: Wiley.

Bickel, P. J. 1965. On some robust estimates of location. *Annals of Mathematical Statistics* 36: 847–858.

Bickel, P. J., and E. L. Lehmann. 1975. Descriptive statistics for nonparametric models II. Location. *Annals of Statistics* 3: 1045–1069.

Breiman, L. 1973. *Statistics: With a View Toward Applications*. Boston: Houghton Mifflin.

Cox, N. J. 1999. gr42: Quantile plots, generalized. *Stata Technical Bulletin* 51: 16–18. Reprinted in *Stata Technical Bulletin Reprints*, vol. 9, pp. 113–116. College Station, TX: Stata Press.

———. 2003. Stata tip 2: Building with floors and ceilings. *Stata Journal* 3: 446–447.

———. 2005. Speaking Stata: The protean quantile plot. *Stata Journal* 5: 442–460.

———. 2012a. Software update: gr42_6: Quantile plots, generalized. *Stata Journal* 12: 167.

———. 2012b. Speaking Stata: Axis practice, or what goes where on a graph. *Stata Journal* 12: 549–561.

Daniell, P. J. 1920. Observations weighted according to order. *American Journal of Mathematics* 42: 222–236.

David, H. A., and H. N. Nagaraja. 2003. *Order Statistics*. 3rd ed. Hoboken, NJ: Wiley.

Davison, A. C., and D. V. Hinkley. 1997. *Bootstrap Methods and Their Application*. Cambridge: Cambridge University Press.

Dixon, W. J. 1960. Simplified estimation from censored normal samples. *Annals of Mathematical Statistics* 31: 385–391.

Dixon, W. J., and F. J. Massey, Jr. 1983. *Introduction to Statistical Analysis*. 4th ed. New York: McGraw–Hill.

Dixon, W. J., and J. W. Tukey. 1968. Approximate behavior of the distribution of Winsorized t (trimming/Winsorization 2). *Technometrics* 10: 83–98.

Dixon, W. J., and K. K. Yuen. 1974. Trimming and Winsorization: A review. *Statistische Hefte* 15: 157–170.

Donoho, D. L., and P. J. Huber. 1983. The notion of breakdown point. In *A Festschrift For Erich L. Lehmann*, ed. P. J. Bickel, K. Doksum, and J. L. Hodges, Jr., 157–184. Belmont, CA: Wadsworth.

Efron, B., and R. J. Tibshirani. 1993. *An Introduction to the Bootstrap*. New York: Chapman & Hall/CRC.

Ender, P. B. 2009. robmean: Trimmed, Winsorized means & Huber 1-step estimator. UCLA: Statistical Consulting Group.
http://www.ats.ucla.edu/stat/stata/ado/analysis.

Erickson, B. H., and T. A. Nosanchuk. 1977. *Understanding Data*. Toronto, Canada: McGraw–Hill.

———. 1992. *Understanding Data*. 2nd ed. Toronto, Canada: University of Toronto Press.

Feigelson, E. D., and G. J. Babu. 2012. *Modern Statistical Methods for Astronomy: With R Applications*. Cambridge: Cambridge University Press.

Flournoy, N. 1993. A conversation with Wilfrid J. Dixon. *Statistical Science* 8: 458–477.

———. 2010. Wilfrid Joseph Dixon, 1915–2008. *Journal of the Royal Statistical Society, Series A* 173: 455–457.

Gauss, C. F. 1816. Bestimmung der Genauigkeit der Beobachtungen. *Zeitschrift für Astronomie und verwandte Wissenschaften* 1: 187–197.

Gordin, M. D. 2004. *A Well-Ordered Thing: Dmitrii Mendeleev and the Shadow of the Periodic Table*. New York: Basic Books.

Hamilton, L. C. 1991. sed4: Resistant normality check and outlier identification. *Stata Technical Bulletin* 3: 15–18. Reprinted in *Stata Technical Bulletin Reprints*, vol. 1, pp. 86–90. College Station, TX: Stata Press.

Hampel, F. R. 1985. The breakdown points of the mean combined with some rejection rules. *Technometrics* 27: 95–107.

———. 1997. Some additional notes on the "Princeton robustness year". In *The Practice of Data Analysis: Essays in Honor of John W. Tukey*, ed. D. R. Brillinger, L. T. Fernholz, and S. Morgenthaler, 133–153. Princeton, NJ: Princeton University Press.

Hampel, F. R., E. M. Ronchetti, P. J. Rousseeuw, and W. A. Stahel. 1986. *Robust Statistics: The Approach Based on Influence Functions*. New York: Wiley.

Harter, H. L. 1974a. The method of least squares and some alternatives—Part I. *International Statistical Review* 42: 147–174.

———. 1974b. The method of least squares and some alternatives—Part II. *International Statistical Review* 42: 235–264 and 282.

Helsel, D. R., and R. M. Hirsch. 1992. *Statistical Methods in Water Resources*. Amsterdam: Elsevier. http://pubs.usgs.gov/twri/twri4a3/.

Hill, M., and W. J. Dixon. 1982. Robustness in real life: A study of clinical laboratory data. *Biometrics* 38: 377–396.

Huber, P. J. 1964. Robust estimation of a location parameter. *Annals of Mathematical Statistics* 35: 73–101.

————. 1972. The 1972 Wald lecture: Robust statistics: A review. *Annals of Mathematical Statistics* 43: 1041–1067.

————. 1981. *Robust Statistics.* New York: Wiley.

Huber, P. J., and E. M. Ronchetti. 2009. *Robust Statistics.* 2nd ed. Hoboken, NJ: Wiley.

Jennrich, R. I. 2007. BMDP and some statistical computing history. *Statistical Computing and Graphics* 18: 17–23.

Jurečková, J., and J. Picek. 2006. *Robust Statistical Methods with R.* Boca Raton, FL: Chapman & Hall/CRC.

Kafadar, K. 2003. John Tukey and robustness. *Statistical Science* 18: 319–331.

Kim, S.-J. 1992. The metrically trimmed mean as a robust estimator of location. *Annals of Statistics* 20: 1534–1547.

Kleiner, B., and T. E. Graedel. 1980. Exploratory data analysis in the geophysical sciences. *Reviews of Geophysics and Space Physics* 18: 699–717.

Lehmann, E. L. 1983. *Theory of Point Estimation.* New York: Wiley.

Mazliak, L. 2012. Poincaré's odds. *Séminaire Poincaré* 16: 173–210.

Miller, R. G. 1986. *Beyond ANOVA: Basics of Applied Statistics.* New York: Wiley. Reprinted, London: Chapman & Hall, 1997.

Mosteller, F., and J. W. Tukey. 1977. *Data Analysis and Regression: A Second Course in Statistics.* Reading, MA: Addison–Wesley.

Olive, D. J. 2001. High breakdown analogs of the trimmed mean. *Statistics & Probability Letters* 51: 87–92.

Pearson, R. K. 2011. *Exploring Data in Engineering, the Sciences, and Medicine.* New York: Oxford University Press.

Poincaré, H. 1896. *Calcul des probabilités.* Paris: Gauthier-Villars.

————. 1912. *Calcul des probabilités.* 2nd ed. Paris: Gauthier-Villars. http://archive.org/details/calculdeprobabil00poinrich.

Reimann, C., P. Filzmoser, R. Garrett, and R. Dutter. 2008. *Statistical Data Analysis Explained: Applied Environmental Statistics with R.* Chichester, UK: Wiley.

Rice, J. A. 2007. *Mathematical Statistics and Data Analysis.* 3rd ed. Belmont, CA: Duxbury.

Rosenberger, J. L., and M. Gasko. 1983. Comparing location estimators: Trimmed means, medians, and trimean. In *Understanding Robust and Exploratory Data Analysis,* ed. D. C. Hoaglin, F. Mosteller, and J. W. Tukey, 297–338. New York: Wiley.

Short, J. 1763. Second paper concerning the parallax of the sun determined from the observations of the late transit of Venus, in which this subject is treated of more at length, and the quantity of the parallax more fully ascertained. *Philosophical Transactions of the Royal Society of London* 53: 300–345.

Siegel, A. F. 1988. *Statistics and Data Analysis: An Introduction*. New York: Wiley.

Sprent, P., and N. C. Smeeton. 2007. *Applied Nonparametric Statistical Methods*. 4th ed. Boca Raton, FL: Chapman & Hall/CRC.

Staudte, R. G., and S. J. Sheather. 1990. *Robust Estimation and Testing*. New York: Wiley.

Stewart, C. A. 1947. P. J. Daniell. *Journal of the London Mathematical Society* 22: 75–80.

Stigler, S. M. 1973. Simon Newcomb, Percy Daniell, and the history of robust estimation 1885–1920. *Journal of the American Statistical Association* 68: 872–879.

———. 1976. The anonymous Professor Gergonne. *Historia Mathematica* 3: 71–74.

———. 1977. Do robust estimators work with real data? *Annals of Statistics* 5: 1055–1098.

———. 2010. The changing history of robustness. *American Statistician* 64: 277–281.

Tilanus, C. B., and G. Rey. 1964. Input-output volume and value predictions for the Netherlands, 1948–1958. *International Economic Review* 5: 34–45.

Tukey, J. W. 1960. A survey of sampling from contaminated distributions. In *Contributions to Probability and Statistics: Essays in Honor of Harold Hotelling*, ed. I. Olkin, S. Ghurye, W. Hoeffding, W. Madow, and H. Mann, 448–485. Stanford, CA: Stanford University Press.

———. 1962. The future of data analysis. *Annals of Mathematical Statistics* 33: 1–67.

———. 1970a. *Exploratory Data Analysis*. Limited preliminary ed. Reading, MA: Addison–Wesley.

———. 1970b. Some further inputs. In *Geostatistics: A Colloquium*, ed. D. F. Merriam, 163–174. New York: Plenum.

———. 1977. *Exploratory Data Analysis*. Reading, MA: Addison–Wesley.

———. 1986. Choosing techniques for the analysis of data. In *The Collected Works of John W. Tukey: Philosophy and Principles of Data Analysis: 1965–1986, Volume IV*, ed. L. V. Jones, 869–874. Monterey, CA: Wadsworth and Brooks/Cole.

Tukey, J. W., and D. H. McLaughlin. 1963. Less vulnerable confidence and significance procedures for location based on a single sample: Trimming/Winsorization 1. *Sankhyā* 25: 331–352.

Turner, G. L'E. 1969. James Short, F.R.S., and his contribution to the construction of reflecting telescopes. *Notes and Records of the Royal Society of London* 24: 91–108.

van Belle, G., L. D. Fisher, P. J. Heagerty, and T. Lumley. 2004. *Biostatistics: A Methodology For the Health Sciences.* 2nd ed. Hoboken, NJ: Wiley.

Venables, W. N., and B. D. Ripley. 2002. *Modern Applied Statistics with S.* 4th ed. New York: Springer.

Wilcox, R. R. 2003. *Applying Contemporary Statistical Techniques.* San Diego: Academic Press.

———. 2009. *Basic Statistics: Understanding Conventional Methods and Modern Insights.* New York: Oxford University Press.

———. 2010. *Fundamentals of Modern Statistical Methods: Substantially Improving Power and Accuracy.* 2nd ed. New York: Springer.

———. 2012a. *Introduction to Robust Estimation and Hypothesis Testing.* 3rd ed. Waltham, MA: Elsevier.

———. 2012b. *Modern Statistics for the Social and Behavioral Sciences: A Practical Introduction.* Boca Raton, FL: CRC Press.

Woolf, H. 1959. *The Transits of Venus: A Study of Eighteenth-century Science.* Princeton, NJ: Princeton University Press.

Wulf, A. 2012. *Chasing Venus: The Race to Measure the Heavens.* New York: Alfred A. Knopf.

12 Vignettes

Percy John Daniell (1889–1946) was born to British parents in Valparaiso, Chile. He was the last publicly declared Senior Wrangler in Mathematics (top student in his year) at Cambridge in 1909. After brief periods in Liverpool and Göttingen, he taught and researched from 1912 to 1923 at the Rice Institute at Houston, Texas, before returning to Britain as Professor of Mathematics at Sheffield. Daniell's contributions, which span a wide range from pure mathematics to applied mathematics and statistics, were surveyed briefly by Stewart (1947) and Stigler (1973) and in much more detail by Aldrich (2007).

Wilfrid Joseph Dixon (1915–2008) was born in Portland, Oregon. He received degrees in mathematics and statistics from Oregon State College, the University of Wisconsin, and Princeton. He was on the faculty at the University of Oklahoma, the University of Oregon, and University of California, Los Angeles, where he was a leader in biostatistics and biomathematics. Dixon's statistical interests were wide ranging, including robust estimation in the presence of outliers, and he collaborated with medical scientists on many projects. With Frank Jones Massey, Jr. (1919–1995), he wrote a major statistics text for nonmathematicians, which was unusual in including material on trimming and Winsorizing in its third and fourth editions (1969, 1983). Beginning in 1961, he led the development of the package that has morphed over its history from BIMED to BIMD to BMD to BMDP. See Flournoy (1993, 2010) and Jennrich (2007) for more details.

Donald Hatch McLaughlin (1941–) earned degrees in mathematics and psychology from Princeton, the University of Pennsylvania, and Carnegie Mellon and taught psychology at Berkeley for six years. Since 1973, he has worked for the American Institutes for Research in Palo Alto, California, and independently as a senior researcher and consultant on many applied projects in education and several other areas.

Dmitrii Ivanovich Mendeleev (1834–1907) was born near Tobolsk in Siberia. He studied and researched in chemistry in St. Petersburg and Heidelberg, quickly rising to professorial rank and establishing St. Petersburg as a major center in chemical research. Mendeleev is best known for his work developing a periodic table of the elements, distinguished not only for providing a classification but also for allowing the prediction of other elements and correcting errors in the measurement of atomic weights. He was, however, much more than an outstanding chemist: "The same individual who composed the periodic system also helped design the highly protectionist Russian tariff of 1891, battled local Spiritualists, created a smokeless gunpowder, attempted Arctic exploration, consulted on oil development in Baku, investigated iron and coal deposits, published art criticism, flew in balloons, introduced the metric system, and much more" (Gordin 2004, xviii). Numerous different transliterations of his name exist.

James Short (1710–1768) was born in Edinburgh and first educated to become a minister, but with inspiration and support from Colin MacLaurin, he became more interested in mathematics and optics and specifically the construction of telescopes. He used metallic specula and succeeded in giving them true parabolic and elliptic shapes. Short adopted telescope-making as his profession, practicing with great success in Edinburgh and then London. He was elected Fellow of the Royal Society and published many of his observations, including his calculation of solar parallax from the 1761 transit of Venus. See also Turner (1969). Note that Short was Scottish, not English as stated by Stigler (1973, 873).

John Wilder Tukey (1915–2000) was born in New Bedford, Massachusetts. He studied chemistry at Brown and mathematics at Princeton and afterward worked at both Princeton and Bell Labs. He was also involved in a great many government projects, consultancies, and committees. He made outstanding contributions to several areas of statistics, including time series, multiple comparisons, robust statistics, and exploratory data analysis. Tukey was extraordinarily energetic and inventive, not least in his use of terminology: he has been credited with inventing the terms "bit", "analysis of variance", "box plot", "data analysis", "hat matrix", "jackknife", "stem-and-leaf plot", "trimming", and "Winsorizing", among many others. He was awarded the U.S. National Medal of Science in 1973. Tukey's direct and indirect influence marks him as one of the greatest statisticians of all time.

Charles P. Winsor (1895–1951) was educated at Harvard as an engineer and then worked for the New England Telephone and Telegraph Company, but his interests shifted to biological research and biostatistics. After further study at Johns Hopkins and Harvard, he held posts at Iowa State College and Johns Hopkins; in between, in the Second World War, he did government work at Princeton. The term "Winsorize" has been attributed to J. W. Tukey but was first used in publications by Dixon (1960).

About the author

Nicholas Cox is a statistically minded geographer at Durham University. He contributes talks, postings, FAQs, and programs to the Stata user community. He has also coauthored 15 commands in official Stata. He was an author of several inserts in the *Stata Technical Bulletin* and is an editor of the *Stata Journal*.